Ergonomics and Diversity, Equity, and Inclusion

Diverse, equitable, and inclusive (DEI) workplaces foster a sense of belonging and support for all employees, helping them feel like integral parts of the team. Yet, many organizations underestimate the significance of ergonomics in creating such environments and fail to recognize how their ergonomics practices promote DEI and improve workplace culture. Why should ergonomics professionals and practitioners pay attention to DEI, and how can DEI professionals and advocates benefit from understanding ergonomics?

This book investigates the symbiotic links between ergonomics and DEI and how workplaces and work practices can be improved through a better understanding of both. This book reveals how to utilize DEI strategies to improve ergonomics and human factors efforts, and how professionals, in turn, can apply their expertise to improve DEI strategies. This book also serves as a resource for those looking for tangible ideas to expand their DEI efforts through the practical utilization of ergonomics, complete with real-world examples. It offers perspectives and practical advice from ergonomics and DEI experts on elements of sustainable ergonomics processes, elements of a successful corporate DEI strategy, and how to combine the two to create a partnership that will improve workspaces, business strategies, and the quality of life for employees and/or end users of products and services designed by engineers. Featuring guidance and discussion points related to how critical diversity is to successful ergonomics programs, this book delivers ergonomics principles that can be used to promote inclusive design and how to enhance and showcase the value of ergonomics to businesses via incorporation with corporate DEI strategies and vice versa.

Ergonomics and Diversity, Equity, and Inclusion: Theory and Practice will be a valuable reference for professionals and practitioners in ergonomics and human factors, business and management, human resource management, industrial and systems engineering, mechanical engineering, and computer science engineering and for those seeking contemporary guidance to inform their DEI policies and practices.

Ergonomics and Diversity, Equity, and Inclusion

Theory and Practice

Bobbie Watts

CRC Press
Taylor & Francis Group
Boca Raton London New York

CRC Press is an imprint of the
Taylor & Francis Group, an **informa** business

Designed cover image: Derick Harris

First edition published 2025
by CRC Press
2385 NW Executive Center Drive, Suite 320, Boca Raton FL 33431

and by CRC Press
4 Park Square, Milton Park, Abingdon, Oxon, OX14 4RN

CRC Press is an imprint of Taylor & Francis Group, LLC

ISBN: 978-1-032-65473-7 (hbk)
ISBN: 978-1-032-65897-1 (pbk)
ISBN: 978-1-032-65898-8 (ebk)

DOI: 10.1201/9781032658988

Typeset in Times
by Deanta Global Publishing Services, Chennai, India

For mama and daddy.

Contents

PART III The Current and Future State of DEI

About the Author

Bobbie Watts, PhD, CPE, is a recognized expert with over 27 years of experience in Industrial Engineering and Ergonomics/Human Factors. She holds a BS in Industrial Engineering from Clemson University, as well as a master's and PhD in Industrial and Systems Engineering from Auburn University. Her career includes key roles such as Corporate Human Factors Engineer at UPS, consultant for USPS and American Express, Corporate Ergonomics Manager at Coca-Cola Refreshments, and Director of Ergonomics for Michelin Americas. In her current role as an Innovation Initiative Leader and Start-up Founder within Michelin, Dr Watts is dedicated to developing assistive technologies that improve mobility freedom for wheelchair users. Driven by a passion for improving the quality of life for workers and end users through ergonomics, Dr Watts focuses on designing environments and technologies that are inclusive, equitable, and accessible to all. As a Riley Diversity Leadership Institute Fellow, she advocates for a more diverse and inclusive approach to ergonomics, ensuring that all individuals, regardless of background or ability, benefit from advancements in workplace design. She also served as President of the IISE Applied Ergonomics Society, where she emphasized the importance of equity and inclusion within the ergonomics community.

Opening Introduction

If you have found your way to this book, I will have to assume you have at least some level of interest in Ergonomics/Human Factors. Perhaps you're also interested in Diversity, Equity, and Inclusion (DEI), or specifically, in DEI as a comprehensive organizational strategy. If you happened upon this book by chance, I'm grateful for the fortunate turn of events that brought you here! Either way, welcome! You are in for what I hope is an insightful exploration as I share what I consider an ode to the fields of Ergonomics/Human Factors and DEI. In these pages, I dive headfirst into the interconnectedness of the two domains, exploring how their combined application (dare I say, "synergy") in workplaces can be strategically managed to drive meaningful change. While the potential for this connection to transform organizations is clear in theory, its practical realization depends on us—for which I hope this book can add to the discourse.

But first, let's level set on what I will be talking about.

DEFINING ERGONOMICS/HUMAN FACTORS

First off, let's examine "Ergonomics/Human Factors." Technically speaking, according to the International Ergonomics Association (IEA, 2003) the term "ergonomics," often used interchangeably with the term "human factors," refers to a scientific discipline focused on understanding the interactions between humans and other elements within a system—with a system being (again, re: IEA) a complex arrangement of interacting elements including humans, machines, environment, and organizational factors. A system can be anything such as—and forgive the simplistic examples—something as simple as me using a pen on paper for the purpose of recording my thoughts, or something more complex, like a business organization. Ergonomics/Human Factors applies theory, principles, data, and methods to design systems that optimize human well-being and overall system performance (note how human is first in that system equation). As a multidisciplinary, user-centered science, Ergonomics/Human Factors professionals, students, specialists, technicians, and those affiliated with the field through practice work to address a wide range of issues and challenges (*note for simplicity, I will refer to the field as "ergonomics" and the worker collective as "ergonomics practitioners" throughout the book*). These include improving physical and mental safety, promoting physical and mental health, and considering cognitive and psychosocial aspects of the human-system-environment interaction, whether work-related or otherwise. The goal is to create systems that are safe, effective, efficient, and sustainable.

To achieve this, ergonomics practitioners must consider a broad spectrum of factors and domains, including:

- **Physical Ergonomics**: A focus on human anatomical, physiological, and biomechanical characteristics in relation to physical activity (i.e., human anatomy, physiology, anthropometrics, biomechanics).

DOI: 10.1201/9781032658988-1

- **Cognitive Ergonomics**: A focus on mental processes, such as perception, memory, reasoning, and motor response, and how they impact interactions between humans and other system elements (i.e., perception, memory, reasoning, and motor response).
- **Organizational Ergonomics**: A focus on optimizing sociotechnical systems, including organizational structures, policies, and processes (i.e., participation, cooperation, environment, sociotechnical systems).

When all of these elements are taken into consideration, ergonomics practitioners can enhance both human well-being *and* system performance, ensuring that systems are not only functional but also aligned with human needs and capabilities.

However, to achieve this effectively, it is essential for our profession to recognize the wide range of diverse needs and abilities, understand how this diversity influences human-system interaction, and ensure that every aspect of the organizational ergonomics processes we implement is designed to inclusively and equitably address those varied needs. Tall order, huh?

DEFINING DIVERSITY, EQUITY, AND INCLUSION

When it comes to defining "Diversity, Equity, and Inclusion", well, that can get more subjective with regards to how the individual elements combine into a collective strategy used within organizations (*and again, for simplicity, I use "DEI" when referring to the collective strategy*). When I'm asked what "DEI" means to me, I like to provide a quick, personal explanation that captures the essence of it. In past presentations on the topic, I've shared something along these lines:

> DEI is putting actions into place to ensure the wide range of our varied experiences and perspectives are considered and included.

This definition, of course, reflects how DEI has evolved into a more comprehensive and holistic concept. It is no longer just about a single characteristic but an amalgamation of the three distinct but interrelated principles. To arrive at my personal explanation/definition, I looked at each of the words individually, using definitions that align with the spirit of the terms (from definitions derived from the Oxford English Dictionary):

- DIVERSITY: The practice or quality of **including or involving people** from a range of different social and ethnic backgrounds and/or different genders, sexual orientations, etc.
- EQUITY: Providing resources appropriate to the environment to **obtain equal outcomes**.
- INCLUSION: The practice or policy of including people who might **otherwise be excluded or marginalized**, such as those who have physical or mental disabilities and members of minority groups.

Of course, the definitions can vary depending on resources and/or context, but for me, I am defining DEI in terms of how I and how I posit that our ergonomics profession approaches it. As such, I usually go on to say with respect to our discipline, DEI, and the design of systems,

> [Ergonomics practitioners] may not see ourselves as designers, but we are. We design programs, processes, tools, equipment, products, experiences... everything! And we do it for EVERYONE so that EVERYONE is safe, working efficiently and able to give of their full self. DEI is in our DNA.

In other words, DEI isn't just an afterthought for us in the ergonomics field; it is central to how we approach designing systems, be they simple or complex, that work for all people.

ERGONOMICS AND DEI: THE CONNECTION

In the book, I will discuss how we in the ergonomics field often struggle to communicate the true depth of our discipline. Sure, all of us in the ergonomics circle may know how important ergonomics is and how it helps the world spin, but let's face it, that is not the perception outside our world. When many people outside of the field think of ergonomics, chances are they will immediately think of "keyboards," "chairs," or "carpal tunnel syndrome." While those topics are part of what we do, we know that ergonomics encompasses so much more. Going back to the earlier statement that ergonomics can enhance both human well-being and system performance, ensuring that systems are not only functional but also aligned with human needs and capabilities, that in itself speaks to the value we can bring to the table in any organization. And it is time we highlight ergonomics as a universal tool that extends far beyond safety—including being a strategic contributor to business outcomes (Mulaomerovic and Wang, 2023).

And one way we can highlight that is through highlighting our connection with DEI. For me, the connection between DEI and ergonomics has always been obvious. I simply can't see one without the other, especially when it comes to making our ergonomics efforts truly effective. But unfortunately, that connection is not readily visible to others—both outside and inside our profession. Here's me hoping we can change that, from the inside out.

In the book, I will go further into that connection, beginning with just how integral DEI (and each individual element) is to ergonomics and how the connection can be further developed to enhance both domains. It is an interesting time to have this conversation. When "DEI" became a key focus for corporate strategies, especially with the increase in focus after the global racial justice movements of 2020, it presented a unique opportunity for ergonomics to be recognized for its potential to contribute to those DEI goals. And now, in 2025, the need for the connection between DEI and ergonomics has only grown stronger.

According to Culture Amp, an employee experience platform, DEI has recently faced negative attention due to misunderstandings, political pressure, and frustration over lackluster results. Despite this, the benefits of diverse, inclusive environments on safety, employee well-being, and overall business performance remain clear. As a result, many companies are still committed to DEI but are shifting toward more integrated, systemic approaches to drive equity, often weaving DEI principles into broader existing organizational programs (CultureAmp, 2024). If there was ever a time to highlight how ergonomics fits seamlessly into these broader, integrated approaches, it is now! By aligning DEI efforts with existing ergonomics processes, companies can leverage ergonomics as a key driver of DEI success—particularly when that success is measured by collaborative metrics.

When ergonomics and DEI come together, the impact can be tremendous—not only due to how ergonomics expertise can strengthen DEI efforts but also how a DEI focus can improve and evolve our organizational ergonomics processes, as well as enhance the public and organizational marketing that our ergonomics profession (in my opinion) desperately needs.

Ultimately, my goal is to bridge and connect these two fields in a way that benefits both, as well as other organizational strategies—including Safety, Operational Efficiency, Employee Experience, and Sustainability. I want to inspire more collaboration between ergonomics practitioners and DEI stewards to create workplaces and systems that are not only efficient and effective but also inclusive, supportive, safe, and accessible to everyone in a way that everyone can truly thrive. This partnership has the potential to improve the experience and outcomes for individuals and organizations alike, and I'm hoping this book will be one of many catalysts for those conversations and changes.

So, What Is This Book Really about?

Glad you asked! In short, this book will explore, through theory and practical examples, how critical diversity is to successful organizational ergonomics processes and how ergonomics principles can promote equitable and inclusive design. It will also offer practical ways to highlight the value of ergonomics to businesses/organizations, especially in building a culture of employee engagement and innovation, through integration with DEI strategies—and vice versa. As such, I hope this book can serve as a valuable reference for ergonomics practitioners, DEI practitioners, and anyone else looking to find ways to develop and design systems and environments to be inclusive and equitable while recognizing and leveraging the diversity around us.

The Benefits Specific to Ergonomics Practitioners

This book was written with ergonomics practitioners in mind—professionally and personally. I want you to know that we, as a community of practitioners, have so much to offer, perhaps more than you may realize. My hope is that this book serves as a reference for you or, at the very least, provides confirmation that you have the expertise and potential to make a profound impact, whether you're a seasoned professional or just starting out.

The work we do as ergonomics practitioners is noble. We improve quality of life by designing work environments that allow workers/end users to be safe, comfortable, and effective. We also have a key role to play in helping organizations create inclusive environments where a diverse workforce can thrive. And this is crucial—because when it comes to sustainable culture change within organizations, inclusion is at the heart of it.

Within these pages, you will find a blend of technical guidance and leadership competencies that are crucial in today's rapidly evolving world. I will delve into how to develop an inclusive mindset—because in ergonomics, it is not just about technical expertise, it is about being able to see the bigger picture, ask the right questions, and pull in the right tools (and people) as needed. It is not just about what we know, but how we apply it, how we engage with others, and how we use our knowledge to drive change.

In addition to technical expertise, leadership competencies are becoming increasingly essential. In our fast-paced, culturally shifting world, it is important to stand out with not only technical skills but also the leadership qualities that will empower diverse teams. Effective leadership includes the ability to coach, support, and develop teams with a focus on empathy, social competencies, and empowerment—areas that are also key to advancing DEI efforts and goals. By nurturing these leadership qualities, you can diversify your career path, which, in turn, can help diversify any organization you are a part of. For some, this could even mean stepping into non-traditional "ergonomics" roles (gasp!) such as organizational leadership roles, further expanding the impact you can have on your organization and beyond.

Beyond career development, this book aims to elevate the ergonomics profession and bring it the recognition it so truly deserves. While I may sometimes over-enthusiastically talk about the brilliance of our field (apologies in advance!), it is time we address the fact that ergonomics often goes underappreciated outside of our immediate circles.

The Benefits Specific to DEI Practitioners

To those serving as DEI stewards and advocates, please recognize that the work you do in this space is equally noble. You strive to inspire and motivate others to embrace diversity while fostering the development of support systems within organizations to ensure that diversity is both respected and leveraged. You serve as a role model for the principle that everyone deserves to be treated with fairness and respect. And as such, you deserve all the support possible to be able to succeed.

You might not fully realize the impact ergonomics can have on your efforts to advance DEI within your organization and drive positive business performance. By exploring the principles and insights in this book—on why and how work systems and environments should be designed with DEI in mind, how DEI can be integrated into ergonomics processes to foster innovation through employee engagement, and how both DEI and ergonomics success measures can demonstrate tangible progress in business and social performance to employees, the C-suite, shareholders, and the public—we can collectively build a strong value proposition.

My hope is that DEI professionals, committees working on DEI initiatives, and other DEI advocates will find practical, actionable ideas to enhance their efforts by collaborating with ergonomics experts within their organizations and applying ergonomics principles to their work. If this book inspires even one person outside the ergonomics field, from a different profession or industry, to take a deeper interest in ergonomics and its intersection with DEI, then the countless hours spent writing, researching, and interviewing others for their perspectives will have been well worth it.

GOALS OF THE BOOK

Now, that brings me to the goals of the book. And I have several.

First, I want this book to be a resource for those who are looking for fresh guidance on why and how to integrate DEI strategies into their work and/or studies. This is especially geared toward those working in industrial or office settings, or those in academia who focus on applying ergonomics principles in industry.

Second, I aim to educate key business stakeholders on the benefits of ergonomics, DEI, and the powerful combination of both. Diversity, Equity, and Inclusion remains a challenge that many organizations are looking to address more effectively, particularly in today's climate. But the goal isn't just to educate the outside world; we also need to educate ourselves within the ergonomics community. We are already engaged in work that promotes inclusivity and equity, and we rely heavily on the diversity of our workforce. Let's take pride in these achievements, highlight our successes, and celebrate the impact we're already making. This book will also encourage us to think bigger—not just to continue the work we're doing, but to enhance it. We can further leverage diversity to drive innovation, design more intentionally for inclusion, expand our focus on psychophysical considerations, and become champions of psychological safety—promoting physical, emotional, and cognitive well-being within organizations.

It is important to note upfront that this book won't provide everything you need to know. That would be an impossible task, as each topic could easily fill an entire book on its own. Instead, the aim is to provide foundational theory, real-world examples of how these concepts are being applied in the field, and thoughts and considerations to ponder on. My hope is that you'll walk away with key takeaways—insights you can return to time and again as you continue to grow and evolve in your work, whether as an ergonomics practitioner, DEI practitioner, or champion of both.

WHY ME?

That's a great question. I was honored to be selected for the Riley Institute at Furman's Diversity Leaders Initiative (DLI), a program that brings together leaders to break down barriers and drive economic progress in South Carolina. Through this experience, I gained valuable insights and tools for leveraging diversity to improve organizational outcomes and create social and economic impact, especially in our local communities. These lessons are woven throughout this book.

However, being a DLI Fellow doesn't make me a DEI expert, nor did I write this book because I consider myself one. The DLI experience helped me develop the skills to see and appreciate the diversity around me and opened my mind to new perspectives. I wrote this book because I believe we all have something to contribute in this space, and I simply have a perspective I want to share with the world.

My perspective is shaped by many aspects of who I am. It is influenced by my experiences and outlook as a Black woman, born on a farm in the southern United States, and as the granddaughter of a sharecropper. I navigated my way to (and through!) Clemson University as a first-generation college student (in my immediate family), then on to Auburn University (twice!), eventually finding my way into a profession I love. It is a profession I carry with me and apply in every aspect of my life, no matter the job title.

My career journey, some of which you will read about in the upcoming chapters, has taken me through many diverse jobs and industries, including manufacturing, services, research and development, utilities, government, and as a consultant for many more. I have been an Industrial Engineer, Human Factors Engineer, Ergonomics Coordinator, Corporate Ergonomist, Ergonomics Director, Principal Ergonomist, Professor, and Innovation Initiative Leader. I have developed and designed products, tools, processes, and experiences, served as the consulting human factors expert on small- and large-scale design initiatives, created and coordinated participatory ergonomics processes for site-level and organizational-level programs, managed teams of ergonomics practitioners across continents, and led the design, development, and business exploration for assistive technology products. Needless to say, my perspective has been shaped by those experiences, including the people—workers, coworkers, students, clients, and consumers—I met along the way.

As such, my perspective is very much rooted in my deep passion for ergonomics and the power this field possesses to truly make a difference. It may sound cliché, but I genuinely believe that. What drew me to ergonomics is the same reason I am passionate about advancing DEI. While people may have different interpretations of "DEI," at its heart, it is all about making everyone feel like they belong—whether in the workplace, office, or any public space. When you feel like you belong, you can give your best, and that has a ripple effect. At the end of the day, it comes down to respect. It is about creating environments—whether in the workplace or beyond—that acknowledge and honor individuals for who they are. And that idea resonates with me on so many levels.

First, it taps into what I believe is the core of humanity: the ability to see and respect one another. While we may not agree on everything, and we don't have to, the key is respecting each person for who they are. Second, creating a world, workplace, or public space where everyone feels they belong does not happen without ergonomics. That is essentially what we do: we identify the best ways to design or redesign spaces, processes, products, and experiences to balance the needs of individuals with the requirements of the system. Third, considering the complexity of people, their diverse needs and how they interact with their environment—whether it is the workplace or beyond—now that is the ultimate puzzle! It is a big one, yes, but just thinking about the ideas, creativity, and innovation that have already been sparked by this challenge (hello, artificial intelligence anyone?!)—and the future possibilities—is incredibly exciting. To be part of that, to help innovate while also helping others, is enough to fuel my passion for a lifetime. And if I know this profession, I am not alone in that passion.

So, yes… I want my passion for ergonomics and DEI to be added to the discourse. I hope that the practical guidance shared within this book helps to break through the noise that is surrounding "DEI," especially as of late. Because it is hard to argue against the practicality of designing "work" such that the system is so balanced that we *all* win. By designing work environments and systems with ergonomics principles that fully encompass the diverse complexity of people in mind, we can create scenarios where everyone benefits, leading to a win-win situation for all employees and the organization as a whole.

So, why me? Why not me? Why not all of us?

HOW IT STARTED

In the beginning stages of the book, I queried several ergonomics professionals; a few I cornered at one of the professional ergonomics conferences I attended to get a feel for how other professionals feel about DEI and our prospective role as ergonomics practitioners and professionals in expanding DEI efforts. Was this just me and my lofty goals of what ergonomics should be? Here are a few points that were shared with me (and throughout the book, you'll find more personal definitions of DEI and insights into the intersectionality between ergonomics and DEI):

Lori Thomas:

> [DEI] is the open dialogue, discussion and inclusion of all individuals regardless of their background, gender, race, religion, physical capability, sexuality… Ergonomics Practitioners can bring better value by understanding where work environments can be more inclusive…

Gary Orr:

> [DEI] is a reminder that decisions should not be made within a homogeneous sphere of people. Equal weight needs to be given to the opinions of people with diverse backgrounds and experiences.

Thomas Sneed:

> [DEI is] giving everyone the same level of attention and service. Not treating everyone the same but treating them in consideration of their wants and needs. It could be gender related, culturally related, physical stature related, psychosocial, etc.

Thomas Rowell:

> It can mean so many things and I think can be very dependent on the context. In the context of ergonomics I think it means we need to be more open minded and aware of the many differences within the human variable of the workplace equation… We ARE

> the profession that is called to build a workplace that fits the worker. By that definition we should look more closely at all of those traits, characteristics, patterns, (call it what you like) that make us different but are part of the human variable.

Ram Maikala:

> ...belonging and be intentional in inclusiveness (not just words but action)... Be intentional and proactive.

Their explanations of "DEI" and its intersection with ergonomics emphasize the need for a holistic approach to inclusivity, one that considers diverse backgrounds, individual needs, and intentional actions to foster belonging. This aligns with our purpose in ergonomics: fitting the job to the person by understanding their unique expectations and needs.

I also had a chance to chat with Gene Kay, a Certified Professional Ergonomist and past president of the Board of Certification in Professional Ergonomics. I wanted to get Gene's perspective on how he views the intersectionality of ergonomics and DEI.

With regards to how DEI fits within the ergonomics profession, Gene views it as integral to what we do as ergonomics practitioners—simply a natural part of our efforts. He pointed out that this is very evident through our training, which encourages us to design/assess for the "entire" population, accommodating a variety of sizes and abilities. He emphasized:

> A good design should cater to diverse backgrounds, not just the majority.

He recognizes, however, that much of the built environment we work within tends to cater to the dominant culture or a dominant portion of the population, which can leave others at a disadvantage. As such, given our natural focus within the ergonomics of designing for a diverse working population, we have a unique opportunity—and responsibility—to promote inclusivity by actively acknowledging the needs of underrepresented groups. This not only makes our designs more effective but also helps everyone feel included in the process. As Gene put it, that is a win-win.

Gene sees the emphasis on inclusivity as particularly critical in the development and execution of our ergonomics processes. Without considering the diverse needs of the workforce in our risk assessments and design strategies, we risk missing the mark entirely. He cautioned:

> If we don't have those varied perspectives, the needs of many employees might not be met efficiently.

And not meeting those needs can lead to a culture that is potentially rife with increased risk and psychological stress, which can lead to injuries and poor operational performance.

This led to a discussion on the importance of representation. Gene stressed the need to consider diverse perspectives in design. If we design with only one type of user in mind—or the "average user" we in ergonomics caution against—we risk excluding many and missing out on valuable insights. Each individual experiences their environment differently, shaped by their physical, mental, and cultural backgrounds. A setup that works for one person could be uncomfortable or unsafe for another, and our risk assessments must account for these differences.

By incorporating diverse perspectives, we also enhance a sense of belonging. As Nicole Fernandes emphasized, belonging is essential to employee engagement. In her Forbes article on "Belonging: The Intersection of DEI and Engagement" (Fernandes, 2021), she highlighted how the concept of "belonging" has become a crucial addition to the DEI lexicon, significantly influencing the success of traditional DEI initiatives and strategies. Given the potential of ergonomics—and by extension, our ergonomics processes—to shape an employee's perception of their work environment, it is easy to see the direct connection between our efforts and the success of DEI strategies. When workers feel their needs are heard and addressed, they're more likely to engage with and support the ergonomics process. An easy way to achieve this is by ensuring input from a broad range of employees.

To understand how ergonomics supports "belonging," we must first connect it to DEI and try meeting employees' basic needs. Maslow's Hierarchy of Needs (McLeod, 2024) outlines foundational needs, from physiological (comfort, safety) to self-actualization (purpose). Ergonomics plays a key role in meeting these needs by adapting work environments to individual capabilities, ensuring comfort, safety, and operational efficiency. This helps employees thrive and grow, fostering long-term well-being and personal development.

Listening to their collective perspectives reinforced my belief in the potential role ergonomics can play in advancing DEI, as well as the importance of leveraging and expanding DEI within our profession to enhance its impact. Ergonomics is essential in promoting DEI goals, particularly in fostering inclusion and a sense of belonging through leveraging diversity and designing for equity. In fact, DEI is deeply embedded in the very foundation of ergonomics. This confirmation strengthened my sense of purpose in creating this book.

When asked how DEI could be better integrated into ergonomics, they emphasized the need to make DEI a central element of our practice. As ergonomics practitioners, this means looking beyond basic physical differences like height and strength and engaging in deeper discussions about user differences—whether physical, psychological, or social. These valuable insights can only be gathered by actively engaging with people from diverse backgrounds and perspectives. In terms of integrating ergonomics more effectively into an organizational strategy like DEI, the consensus was clear: we must continue to prioritize designing for diversity and align our efforts more closely with DEI objectives. With this in mind, the book is structured into three distinct parts to guide the exploration of these interconnected goals.

HOW THE BOOK IS STRUCTURED

In writing this book, I wanted to make sure it wasn't just my perspective that was shared. So, in true "diversity of thought" fashion, I reached out to colleagues in both the ergonomics and DEI fields, from industry and academia, to get their insights.

Their perspectives, including their views and definitions of DEI, personal experiences, and recommendations, are woven throughout the book. I hope this approach not only enriches the content but helps to ensure a broad range of voices are represented.

The focus of the book is heavily on applied ergonomics and DEI within industry. While there's some discussion about digital platforms, the scope of the digital world is so vast that addressing it fully would have made the book much longer. However, the principles and concepts presented are designed to be widely applicable, regardless of the industry.

Although the book offers practical advice, it is important to note that, like in ergonomics, it is impossible to provide a one-size-fits-all checklist for every situation you might encounter. There is a reason a common saying within ergonomics is "it depends"—every situation will be different and will need to be approached differently. Ergonomics, at its core, is about assessing each situation and using knowledge and skills to find multiple potential solutions. Instead, the goal is to offer principles and actionable guidance that can be adapted to specific situations. And because effective solutions must be developed through ongoing engagement with the local workforce, much of the advice centers around fostering that participatory process. Where applicable, I also provide resources (though not exhaustive) that can be useful for specific challenges or at least point you in the right direction.

The flow of the book is structured as follows:

Part I begins with the intersection of ergonomics and DEI in action. I explore how these two concepts are intertwined, from the history of ergonomics as a diverse profession to how we engage with diverse workforces using participatory approaches. I discuss how diversity is embedded in ergonomics, from risk assessments to compliance, and offer real-world examples, including insights from Anthony Banks, a seasoned ergonomics professional, who talks about the importance of diversity in participatory ergonomics and its role in shaping inclusive mindsets from the start of the design process.

Part II focuses on designing for diversity. This section delves into designing for physical diversity, such as differences in gender, age, and cognitive abilities. But it also highlights the importance of accounting for psychosocial factors, such as racial and cultural diversity, and how neglecting these aspects can harm psychological safety in the workplace. The chapters examine how the evolving workplace, driven by technological advancements and changing demographics, must account for all types of diversity. We hear from experts like Rachel Michael to discuss the importance of focusing ergonomics efforts in work design and designing for diversity; Anuja Patil and Andrea Porter, who share their experiences with PPE design for women; and Ann Hall and Lindsay Hawkins, who discuss the impact of not considering gender diversity on psychological safety. James Rethaber joins in to discuss using the Physical Demands Analysis process to identify and prioritize diversity gaps. We also hear from Gary Downey and Victoria Ballard on designing for people with physical and neurodiverse conditions, including discussions on Universal and Inclusive Design principles and a personal Return To Work story from Kevin Doyle on a growing health issue. Gary Allread, Scott Weede, and LaShawn Nevins also join to discuss the challenges and benefits of designing for multiple generations in the workplace.

Part III shifts the focus to how organizations are incorporating DEI into their business strategies and highlights the critical role ergonomics plays in this integration. This section covers the history of DEI strategies, methods for measuring progress, and the mutual benefits of combining DEI and ergonomics efforts. It also addresses the current political challenges surrounding DEI, which are impacting corporate DEI initiatives, and explores how aligning ergonomics objectives, goals, and metrics with DEI strategies can secure long-term support for both areas. Diversity, equity, and inclusion experts Tiffane Davis, Jourdan Saunders, and Marenda Caldwell join me in this section to share their insights on how organizations are navigating these challenges.

So, tuck in and read on to see how we—whether ergonomics practitioners, DEI practitioners, or other stakeholders interested in creating more inclusive and equitable organizations that support and leverages the diversity (current and to come) to improve overall safety, employee well-being, operations, quality, sustainability, and employee experience—can take the opportunities presented within the book to:

- Expand our influence by integrating DEI considerations with ergonomics principles, fostering an inclusive, safer, and more supportive workplace.
- Showcase our value by demonstrating how our ability to merge DEI and innovative thinking in ergonomics creates widespread organizational benefits and highlights leadership skills.
- Secure sustainable support by aligning our ergonomics objectives and metrics with DEI and innovation strategies, positioning ourselves as essential drivers of organizational success.

REFERENCES

CultureAmp. (2024). *Workplace DEI report.* CultureAmp.

Fernandes, N. (2021, December 22). Belonging: The intersection of DEI and engagement. *Forbes.* https://www.forbes.com/sites/nicolefernandes/2021/12/22/belonging-the-intersection-of-dei-and-engagement/

International Ergonomics Association. (2003). *IEA triennial report: 2000–2003.* IEA Press.

McLeod, S. (2024, January 24). *Maslow's hierarchy of needs.* Simply Psychology. https://www.simplypsychology.org/maslow.html

Mulaomerovic, E., & Wang, E. M. Y. (2023). Challenges and opportunities for human factors/ergonomics as a strategic partner for business managers: In-depth research of experts' visions. *Sustainability, 15*(4), 3078. https://doi.org/10.3390/su15043078

Part I

DEI In Ergonomics

Let me share a little secret with you: when I started my career fresh out of Clemson University with a degree in Industrial Engineering, I was completely clueless about what lay ahead. Clueless! Sure, I had the academic knowledge, but stepping into a full-time paid position was a whole different ball game.

My initial position was as a plant industrial engineer, where I performed your traditional industrial engineering tasks like doing time studies (with an analog watch!), creating cost routings (chasing down numbers from every department, rather), and physically mapping out the facility equipped with a tape measuring wheel, X-acto knife, and tape (no AutoCAD for me at that time!). Altogether, these activities laid a sturdy foundation for a newcomer in the field. In addition to these typical IE tasks, however, I was also assigned to serve on an employee-led ergonomics committee. When I say "employee-led," I mean just that—the employees led the team's activities, requests for funding, plans for implementing solutions, and so on. My role as a management representative was to participate in and support these efforts, helping to anticipate and clear any roadblocks that may have come about. Out of all the things I did as a new IE, being a part of that ergonomics team turned out to be the best learning experience I could have asked for.

Here I was, an engineer tasked with improving processes, and what I quickly realized was that the key to making real improvements wasn't just in my technical (and knife) skills. The key was listening to the operators, the people who actually did the work day in and day out. They often held insights into what needed to change to enhance safety, boost productivity, and improve quality.

That experience opened my eyes to the true spirit of my role as a facilitator of process improvement, and it wasn't about swooping in and magically fixing things.

DOI: 10.1201/9781032658988-2

It was about listening, learning, and then applying *that* knowledge. It was the first time I truly understood the power of empathy in engineering. And that is when my fascination with ergonomics really took hold. While I had fallen in love with the theoretical side of the field during my Human Factors course at Clemson, seeing it in action—seeing how understanding the human body, its physiology, biomechanics, and cognitive processes could directly impact how people work—that was the game-changer. I saw that when we worked together and put the worker/user at the forefront, we could make work safer and more productive. Simple as that.

I fell in love with how inclusive and versatile ergonomics could be. In how it is a discipline that anyone can apply and how it thrives on diversity, leveraging different perspectives and experiences to create solutions. And when ergonomics is approached collectively and inclusively, it brings out the best in all of us. Being able to learn from so many others and to impact their lives felt like a "calling" to me. It felt like I had a true sense of purpose. With ergonomics, I finally felt like I had found my home.

WHAT'S TO COME?

In the opening introduction, I laid out the definitions of ergonomics and DEI and how they naturally complement each other. In Part I, I want to delve deeper into how DEI seamlessly and practically integrates into our "ergo world"—how it enhances our profession and drives meaningful change in the workplace. I also want to explore opportunities to further leverage diversity to demonstrate the value we bring to our organizations, while enhancing our ergonomics processes, advancing our profession, and supporting our leadership development as ergonomics practitioners.

In Chapter 1, "Participatory Ergonomics and DEI," I will begin the discussion with an ode to the participatory approach to ergonomics, expounding on the benefits of engagement of all stakeholders, particularly operators, in the design of work systems. The chapter emphasizes the importance of understanding and valuing the diverse experiences and perspectives of individuals in ergonomics and how tapping into these varied perspectives, ergonomics practitioners can enable the development of more effective and inclusive solutions. Additionally, the chapter will delve into guidance on structuring cross-functional ergonomics teams to maximize employee involvement and engagement.

In Chapter 2, "DEI in the Ergonomics Lexicon," I will explore how diversity, equity, and inclusion already benefit from our efforts, while offering suggestions on how we can better showcase this work to emphasize the value we bring to others, including expanding our "ergonomics lexicon." The chapter also emphasizes the need to shift public perceptions of ergonomics, which is often seen solely in terms of injury prevention and human error. It advocates for more proactive communication within the ergonomics community to highlight the often-overlooked contributions, such as improvements to business performance and fostering inclusivity.

In Chapter 3, "DEI in Risk Assessments," I will discuss the importance of ensuring DEI considerations are captured in the risk management aspects of the ergonomics process. It will include a discussion on the need to account for workforce diversity, such as gender, age, physical abilities, and cultural backgrounds, when selecting and implementing ergonomics tools and assessments.

Chapter 4, "DEI and Solution Development" explores the approach to innovation within organizations, highlighting how leveraging the diversity of the workforce can drive process improvements as well as business level success. The Design Thinking user-centered approach, as well as several strategies and activities for nurturing creativity and innovation, are reviewed. The chapter ends with discussions about leveraging diversity to find inspiration for solution development and innovation in uncommon places.

And the final chapter in Part I, Chapter 5, "DEI and Regulatory Compliance," reviews the role of ergonomics in supporting DEI initiatives by aligning with global regulations for workplace and product design. The chapter also addresses the potential challenges with implementing standards and regulations and suggests strategies ergonomics practitioners can employ to help organizations overcome these challenges.

1 Participatory Ergonomics and DEI

"Just how big are these horse mats?"

Not a typical question you'd think would come up in an ergonomics kaizen event for a challenge with a technical process problem, but it was the question that was being asked by me. How did a question like that come up, you may ask? Well, during the start of the event, I wanted to find out something interesting about everyone participating using one of the icebreaker tools (shared later in this chapter). It was through that simple icebreaker that we, well, "broke the ice" and learned more about everyone and what was important to them. What makes them tick. Learning and being actually interested in the varying life experiences in the room can not only open people up to share and speak up, but it can also lead to solutions.

So, back to the "horse mats." It wasn't that we were focused on farm living in the kaizen, but the similarities in the motions involved with hauling horse mats—which can be bulky and weigh over 45 kgs, making them very awkward to maneuver—were very similar to the issue at hand of manipulating large, bulky material. So, it led us down a path of researching tools used in farming to possibly be repurposed for the challenge at hand. It is not to say that the thought of horse matting wouldn't have come up if we didn't do the ice breaker, but there's a good chance the idea could have stayed buried if we didn't tap into the participants' varying life experiences.

Why is that important?

First, let's start with the basics of what it means to have an ergonomics process. As an ergonomics practitioner working within industry, you are in some capacity, tasked with supporting the identification, control, and management of risks that can contribute to injuries, operational inefficiencies, poor employee well-being, or any combination thereof. What that looks like can vary, meaning you're either supporting or facilitating the overall ergonomics process, including efforts and activities designed to improve workplace conditions, optimize workflow, or generally solve challenges that impede within the shop area, facility, office, or company. Where you're doing these activities can vary as well, with ergonomics practitioners working in capacities within the scope of "safety," "health," "human resources", "engineering", "risk management," etc. Which, quite frankly, the diversity in how and where we work within organizations makes sense given our broad reach and intersectionality. Regardless of where we sit (though, there is a strong argument in Part II that ergonomics sits primarily in the purview of work design decisions), the evolution of the process steps associated with "ergonomics programs" has been a part of our ergonomics management toolbox for decades—with many organizations publishing guides and primers for industries, much like the National Institute for Occupational Safety and Health (NIOSH)'s elements involved in sustainable ergonomics programs

DOI: 10.1201/9781032658988-3

(CDC, NIOSH, 2024), Occupational Safety and Health Administration (OSHA)'s Ergonomics Program Management Guidelines For Meatpacking Plants (OSHA, n.d.), or Canadian Centre for Occupational Health and Safety (CCOHS)'s process for implementing ergonomics regulatory requirements (Canada.ca, n.d.) (calibrated based on the organizations' (and specific industries') needs), all involving guides around: leadership commitment, employee participation, training and communication, worksite risk analysis, hazard prevention and control, injury management (given the strong ties to safety management), and overall program management for continuous improvement.

Cultivated in Japan and born out of quality and production control tools that rely heavily on engaging the work experts (operators) in process improvement, Participatory Ergonomics was developed as a macroergonomics approach to more effectively manage the systematic elements of an ergonomics program through the democratization of ergonomics and the process (Imada, 2011). Participatory approaches have been a development within macroeconomics circles since the 1970s, with variations being implemented in several different industries and settings such as community development and public policy (Duraiappah et al., 2005; Sandel & Wiliams, 2024). The act of engagement (in the form of true bottom-up grassroots active participation as opposed to top-down or tokenistic consultation) is why participatory approaches have been effective in achieving sustainable results, as the engagement can promote a sense of ownership and pride. This was evident in an assessment of varied participatory planning approaches utilized in the upgrading of low-income settlements in Brazil (Terdoo, 2024). Participatory Ergonomics follows the same theory of collective engagement in the overall development or improvement of a system.

In technical terms, Participatory Ergonomics is a collaborative method aimed at identifying and addressing workplace hazards or improving opportunities, with a particular focus on preventing musculoskeletal disorders. As outlined by European Union (EU) guidelines, this approach involves engaging workers, supervisors, and other stakeholders in teams to improve working conditions, safety, productivity, quality, morale, and comfort. Team members, having been provided fundamental ergonomics expertise, collaborate to analyze issues and propose solutions that enhance work performance, comfort, and safety, contributing their knowledge, experiences, and ideas. Through this process, they share ownership and responsibility for the success of any design or redesign efforts. The Participatory Ergonomics approach, as opposed to an approach solely reliant on ergonomics experts (internal or external), has over the years been the most implemented ergonomics management tool for organizations given its integration with organizational design and management factors (Candra et al., 2023) and its adaptability to an organization's size, resources, and culture.

Although it can take different forms, Participatory Ergonomics is typically carried out through the structured management of employee-driven teams that lead the efforts within the ergonomics program. According to NIOSH, a Participatory Ergonomics team consists of individuals who work together to design or redesign processes or products that affect their work environment. This definition is widely accepted, with the EU offering a similar description, highlighting that these teams involve workers directly in designing solutions to reduce risks.

I began this book with a focus on the Participatory Ergonomics approach for a reason. To me, it is the cornerstone of our work as ergonomics practitioners. Honestly, I see this approach as the key to our success (or secret weapon, if you will) because it taps into the diverse perspectives of those around us—whether they are from the shop floor, the office, or even our personal lives. It is this blend of expertise and experience that can lead to the most creative and effective solutions. Without the involvement of workers and end-users, our efforts as ergonomics practitioners are often inefficient, if not entirely ineffective.

A major strength of Participatory Ergonomics is the direct involvement of the end users or employees, who provide invaluable insights into their specific work environments and their needs and preferences. Engagement in these approaches, particularly those that encourage direct participation (where employees have some influence over decisions regarding work and workplace changes impacting them), helps to ensure that the solutions developed are practical, effective, and tailored to the unique demands of the employees' roles. This approach not only empowers employees but also fosters a sense of ownership and accountability in maintaining and improving their work environment.

The Participatory Ergonomics approach is diverse in and of itself—having been adapted to a variety of industries and settings. But at its core, it is about drawing on the experiences, knowledge, and perspectives of those involved in doing the work or interacting with the process or product. And if those experiences and perceptions are varied, then it would be in our best interest as stewards of the ergonomics process to acknowledge, understand, and incorporate those varied experiences and perceptions throughout every phase of our ergonomics programs. With a diverse team, we are more likely to have a collection of varied skills, experiences, and perspectives that can be used to enhance the creativity and problem-solving abilities of the team members, generate more ideas, explore more alternatives, and find better solutions. Additionally, it can help us better vet our ideas.

From an ergonomics process perspective, it is in your best interest to not only have a diverse team but also be intentional about taking the time to explore that diversity. For example, taking time to learn about each member and who they are both inside and outside of work can be easily done by setting aside time in every meeting/event just to talk and learn about each other. Or, in the case of the kaizen event I spoke about earlier, embed it into any icebreaker agenda. Doing this will help others to learn what makes them tick and what their interests are. Exploring that diversity not only breaks the ice and gets people comfortable, but it can also lead to solutions.

In that example, during the kaizen event, we learned that two of our participants were ardent ranchers. By taking the time to just talk, laugh, and learn about one another, they eventually felt comfortable enough to talk about some similar challenges they faced when moving and handling horse matting, which are thick rubber mats that provide cushioning and traction for the horses. They shared insights into specialized tools and techniques used on their ranches, including a tractor-mounted device designed for lifting and moving the heavy mats safely. This sparked our curiosity and led the kaizen team to explore a ranching tool as a possible solution for the material handling challenge we were faced with. The team contacted a supplier, arranged a demonstration, and ultimately integrated the tool into their process, resulting in a significant reduction in excessive repetitive force, along with improvements in efficiency and process quality. Sometimes the simplest and most effective solutions can come from taking the time to learn from one another.

To provide further context on the benefits of participatory ergonomics, I reached out to my colleague, Anthony Banks. Anthony is a Certified Professional Ergonomist with over 25 years of experience across various fields, including safety engineering, industrial hygiene, design engineering, lean manufacturing, and ergonomics. He has provided technical and programmatic leadership to numerous Fortune 500 companies and government organizations.

His vast and diverse experience was the reason why I wanted to tap into his insights on the significance of diversity in creating and maintaining effective ergonomics programs and processes, especially his experience in the development of organization-wide Participatory Ergonomics programs. As an initial statement, Anthony emphasized that ergonomics efforts and DEI efforts are deeply interconnected as they rely on each other for success. To him, a crucial factor in maintaining effective workplace ergonomics lies in the democratization of its application, which can only happen in an equitable and inclusive environment. To achieve this effectively, he said it is essential for us as ergonomics practitioners to leverage the diversity within the workforce as this enables us to fully unlock the benefits of ergonomics. Inclusivity is not just beneficial but necessary for any ergonomics program to reach its full potential.

To break down the connection between ergonomics and DEI even further, Anthony further described how the thoughtful integration of ergonomics principles in design in essence serves as the "equity" component within the DEI framework. It is through our ergonomics practices that we achieve and enhance inclusion in our diverse work environments. He added:

> If equity can be defined as "ensuring that access, resources and opportunities are provided for all to succeed and grow, especially for those who are underrepresented and have been historically disadvantaged," then it should be clear how, in practice, this concept aligns with one of the bedrock facilitating practices of ergonomics.

To that end, there is no DEI without ergonomics.

When it comes to Participatory Ergonomics—which he refers to as "participative ergonomics"—Anthony further solidified the benefits of actively engaging workers in identifying improvements and implementing changes in the workplace to enhance productivity and safety. From his experience, he has seen how it is the inclusive environment or a participative ergonomics team that allows team members to draw on their unique experiences, making solution development in this environment both an "art" and a "science." By that, he meant that solutions to nuanced problems are derived in many ways, taking on many forms due to the diversity of backgrounds. And it is this diversity that becomes a catalyst for ideas. He further shared:

> Diversity of backgrounds in an inclusive environment is a tremendous springboard for creativity. Team members routinely draw on their unique backgrounds to provide the approach or inspiration for solutions which enhance processes and reduce the risk of

> musculoskeletal injury. This could be leveraging concepts they learned to make work easier growing up on a farm, concepts from their background as an aircraft mechanic during military service, modifying techniques and supports learned from carrying an instrument while playing ceremonial music in their non-U.S. church traditions, or approaches used to make catching and transporting fish to the local market easier that they learned when growing up in the absence of modern material handling or transportation options, to name a few.

He sees leveraging this aspect of diversity as a superpower in engaged ergonomics. However, it is only possible, he added, if there is a truly inclusive environment present, an environment where team members and participants feel they can truly bring their selves wholly to the task and that their contributions will be valued, as opposed to marginalized or "otherized".

But beyond enhancing solution development, both in quantity and quality, Anthony has seen how Participatory Ergonomics also actively promotes equity by ensuring that workplace designs accommodate a broader range of individuals, enhancing inclusion for those who may have previously faced barriers. He sees this approach as ideally suited to enhance equity and foster inclusion, which can lead to sustained diversity. While we as a profession aim to "design for the most" whenever possible, this approach and mindset enables us to extend those boundaries beyond the typical 90% range (e.g., 5th percentile for females to 95th percentile for males). By focusing on specific tasks and engaging groups of workers, teams can ensure that the design process includes those who may fall outside this central range, demonstrating the feasibility of accommodating a wider variety of users. He stressed that this not only allows current workers to perform their tasks with reduced injury risk and improved efficiency, but that it also opens the door for individuals who might not have previously been able to do the job effectively due to limitations related to strength, body size, visual acuity, or cognitive biases. As he shared:

> Many challenging roles serve as "gateway" positions in industries, leading to promotions and professional growth. Participative Ergonomics can broaden access to these opportunities, thereby enhancing advancement prospects for a more diverse range of individuals.

As far as my personal journey, I like to say that I "cut my ergo teeth" with Participatory Ergonomics. Getting thrown into an ergonomics team, driven mainly by employees, was my first real exposure to ergonomics in industry. And honestly, it shaped my professional development and career advancement in ways that I could not have imagined otherwise. As I mentioned earlier, it allowed me to better comprehend the challenges that employees encountered in their daily work and how their own ideas were often the best solutions to address those challenges. Just like Anthony mentioned in his experiences with participative ergonomics teams, I discovered that the collaborative environment I encountered during my first professional role as an Industrial Engineer taught me that ergonomics is far more effective and sustainable when everyone has a voice. It is especially impactful when the environment encourages people to be their true selves and share their diverse life experiences. Plus, recognizing and valuing each person's input can really boost morale and foster a sense of ownership over the solutions we create together (I like to call it "leaving a legacy"). So, in the end, it is not just about making the workplace safer; it is about building a community where everyone feels heard and appreciated. And let's be honest, who doesn't want to feel like that everywhere they go?

GETTING THE MOST OUT OF YOUR PARTICIPATORY ERGONOMICS TEAMS

Creativity thrives on the diversity of a team's experiences. The greater the variety of backgrounds and perspectives, the greater the potential for generating innovative ideas. But how do we form such a team? A cross-functional collaborative team inherently embodies diversity. Specifically, a cross-functional collaborative ergonomics team has a broad spectrum of knowledge, skills, and experiences among its members, paving the way for an impactful approach to ergonomics.

Referring back to my conversation with Anthony, his suggestion for a well-rounded Participatory Ergonomics approach, and by extension the team make-up itself, usually relies on teams made up of multi-functional capabilities. He noted that these teams can either be formally assembled and trained as a long-term resource for solving challenges (such as a standing ergonomics team that meets at some frequency), or they can be ad-hoc groups created to tackle specific issues within a set timeframe (such as for kaizen events—more on that in a minute). In either scenario, he noted that the "functional diversity" of the team is crucial. This means including stakeholders who can identify problems based on their task perspectives, those who can clarify how the task is *actually* performed (a key distinction between "work as prescribed versus work as done"), and others who can provide insights on which solutions are likely to be accepted or implemented. Lastly, he added, it is important to have stakeholders who can make final decisions on changes, assess the impact of proposed adjustments on the entire process, and possess the technical skills or connections necessary to implement those changes.

More Guidance for Ergo Teams / Kaizen Events

Now that your 'Avengers' have been assembled (or your 'Justice League' if you're more inclined that way), level setting on the goals and objectives for the team or kaizen event is vital for maximizing the potential of its diverse members. Align expectations by:

- ✓ Defining clear, shared deliverables before team meeting / the event.
- ✓ Clarifying each member's tasks and contributions.
- ✓ Agreeing on an agenda, timeline, and key milestones (e.g., action item reviews).
- ✓ Setting norms for updates and decision-making.
- ✓ Establishing guidelines for respect, inclusivity, and active participation.
- ✓ Setting progress metric(s) relevant to the team's/event's objectives (e.g., complete 50% of task assessments by year end; identify 20 potential solutions during the event).
- ✓ Involving everyone in setting expectations to foster commitment and ownership of tasks.

Kaizen Events in Ergonomics

In lean manufacturing, the word "kaizen" means "good change." Developed and popularized by Japanese organizational theorist and management consultant Masaaki Imai, kaizen is a continuous improvement approach used to improve processes through the participation of employees (Feldman, 2023).

Kaizen events are, without a doubt, my go-to approach for solution development as an ergonomics practitioner. I love how they bring people together to tackle a single issue, igniting those creative juices pulled from diverse experiences, leading to actionable plans that come with built-in accountability. Plus, it is a fantastic way to spread ergonomics knowledge while having fun! And if you're not having fun in your kaizen events, then you're absolutely doing it wrong!

Now, the structure and specific steps of a kaizen can differ from person to person; we all pick up our own little tips and tricks along the way. But the essence of a kaizen event is to harness that participatory spirit and turn it into concrete actions aimed at enhancing safety and operational efficiency. Done right, the energy garnered during an ergonomics-focused kaizen event can be infectious! And you might find that you've naturally created "ergo champions" along the way.

At the end of the day, with an effectively facilitated ergonomics-focused kaizen that taps into the talents of the workforce, not only do you walk away with a solid plan to address ergonomics-related concerns while building genuine buy-in from employees. But you also foster a sense of camaraderie among team members that can go beyond the physical improvements of the workspace and into promoting a collaborative and empathetic environment that supports inclusivity. It is truly a win-win—improving processes with creative, innovative thinking while creating lasting connections. What's not to love? (See Chapter 4 to learn more about holding productive kaizen events for innovative solution development).

Gushing about kaizen events aside, having a plan to facilitate the kaizen or even just coordinating dedicated standing ergonomics team meetings begins and ends with a good team. Based on my experience, the make-up of any ergonomics team, especially one participating in events like kaizens for ergonomics solutions, is crucial. Such a team should ideally include (Table 1.1):

- Operators (Subject Matter Experts, or SMEs).
- Management (Supervisors, Department Managers, Plant Management).
- Technical and Maintenance Personnel.
- Medical, Physical Therapy and Safety Experts.
- Engineers (Industrial Engineering/Process).
- Others to be interviewed and engaged.

Each member brings a unique perspective and expertise, contributing to a well-rounded team capable of developing effective ergonomics solutions.

TABLE 1.1
Recommended Ergonomics Team or Kaizen Event Participant List

Role	Expertise	Contribution
Operators (subject matter experts)	Subject-matter expertise and knowledge about the work process, machinery, tools, etc.; lived experience and perspectives	Identifying issues; identifying root causes for issues; validation of solutions; peer buy-in (influence)
Medical/PT/Safety	The relevant medical kinesiology and safety expertise	Assistance with identifying issues and root causes; validation of solutions from a medical/PT/safety process perspective; compliance with standards and regulations; promotion of injury prevention and wellness culture
Technical/Maintenance	Experience and technical knowledge of the equipment, machines, and tools; awareness and visibility of other potential technical solutions	Validation of solutions from a technical process perspective; also identify potential issues for maintenance activities (for event or the proposed solution)
Engineers (industrial engineers/progress)	Experience and technical knowledge of the process; skills designing, modifying, or improving the process with respect to ergonomics, productivity, and quality	Validation of solutions from a process/work standard perspective; guidance regarding the implementation (impact on work standards, head count, etc.)
Interview and engage with others	Additional feedback on the issues and solutions from the perspective of the workers, managers, customers, or other stakeholders not immediately part of the team/event	Identifying issues; identifying root causes for issues; validation of solutions; peer buy-in (influence)

Intentionally selecting team members from various roles will help ensure diverse perspectives and unique needs related to operational roles are captured. This diversity not only amplifies the team's ability to solve problems but also helps ensure that solutions are robust, effective, and tailored to meet the needs of all stakeholders involved. It enables the team to recognize and interpret risks from diverse perspectives and tackle challenges from several angles, thus making it more effective at developing a multitude of quality and efficient solutions. It is also the collaboration within the diverse team that encourages open dialogue and the exchange of ideas, which can instill a sense of ownership and buy-in among team members to foster a safer and more efficient work environment. This diverse team can ultimately promote a culture that values inclusivity, creativity, and continuous improvement in safety and ergonomics practices.

Expanding on the importance of diversity in team composition goes beyond just selecting members from various operational roles; it involves actively seeking out individuals with diverse backgrounds and experiences when possible—such as diversity in tenure/seniority, age, gender, or cultural backgrounds—to ensure those perspectives are embedded in discussions within the team and in conversations with employees not taking part in the event. Building a team with a broad range of viewpoints and problem-solving approaches—or one that is open to actively seeking out those perspectives—helps ensure a more well-rounded approach to addressing challenges. This diverse input enables the team to consider problems from a holistic perspective. As Anthony highlights, this might involve examining not just the technical risks, but also how cultural or group dynamics could impact a worker's approach to a task, as well as the often-overlooked psychosocial factors at play. By doing so, the team can develop more comprehensive solutions that address both the root causes and the contributing factors of the problem.

Another diverse element that could benefit the dynamic of the team would be varied leadership styles. For instance, including employees recognized as natural leaders among their peers can help garner support and buy-in from those not directly involved in the team/event. This leadership endorsement can significantly enhance the implementation and acceptance of solutions derived from the event, even extending its influence beyond the core team. It is also beneficial to include individuals known as "challengers"—people who instinctively play devil's advocate and actively challenge assumptions and ideas (trust me, every work environment has them!). While having a challenger on the team may seem unconventional and counterintuitive, their role can serve as a valuable exercise during solution development events. By questioning assumptions and probing for weaknesses in proposed solutions, challengers can prevent groupthink and surface overlooked issues, ultimately strengthening the robustness of the solutions. However, it is essential to manage challengers carefully to prevent them from derailing the event. This can be achieved by setting clear expectations for their role, encouraging constructive criticism, and ensuring that challenges are aimed at improving ideas rather than dismissing them outright.

GETTING ENGAGEMENT IN A DIVERSE GROUP

Diversity is a powerful asset in problem-solving, bringing together various perspectives, skills, and knowledge. However, simply having a diverse team is not sufficient unless you actively leverage the opportunity to learn from one another and appreciate each person's unique contributions. Therefore, it is crucial to intentionally invest in getting to know your team members better, both on a professional and personal level.

Creating a culture where team members build deeper connections fosters trust, rapport, and collaboration. This deeper understanding also provides valuable insights into their motivations, challenges, preferences, and goals, which can greatly influence your role as a facilitator—and let's face it, that is a vital soft skill to master for any situation! Furthermore, uncovering hidden talents and perspectives within the team can lead to innovative solutions.

One effective approach is to incorporate informal conversations, icebreakers, or other fun, engaging activities into your meetings or events that can help you discover more about each other's hobbies, interests, values, and goals. These interactions allow team members to share more about their hobbies, interests, values, and aspirations. Whether through simple questions, games, or shared experiences like team outings or learning new skills together, the goal is to create a relaxed and inclusive environment where everyone feels comfortable sharing and learning from each other. And who knows, maybe someone's hobby or outside-of-work interest may just be the key solution to a problem!

While some may grumble about the obligatory "introductions" at the start of meetings or events, I have come to appreciate this phase as a chance to build connections. When I lead an event, I eagerly seek out engaging icebreakers to establish a fun atmosphere—because if it is not fun, why do it, eh? These activities not only lighten the mood but also set the tone that the event will be light and interactive. While a quick fix like asking participants to "share an interesting fact about themselves" can work in a pinch, here are other creative icebreakers that I personally find cool:

- Home Screen Picture: With this icebreaker, participants are asked to share, in 30 seconds or less, the reasoning behind the picture on their phone's home screen.
- Roll the Dice: For this activity, you'll need a pair of dice (or a set). You'll assign a question for each number rolled (e.g., for #1: If you won the lottery, what would you do first?; for #2: If you could be paid to teach a subject, what would it be?; and so on) and then have the participants take their chances on the question they get (or for a little razzle dazzle, set it so that the person rolls the dice for the question of the next person).
- Two Truths and a Lie: this one works well for smaller teams that have been working together for a while. It is an exercise where each individual writes out two interesting facts about themselves and one probable lie, and the other participants try to guess which one is the lie. Be sure to allow time for participants to be able to share more details on their experiences. As for me, see if you can guess the lie: 1) I donated a kidney to a stranger; 2) I was once an Atlanta Falcons cheerleader; or 3) I am a 3rd degree black belt in taekwondo.

Be sure to take detailed notes on what participants share during the event and integrate these insights as the activity progresses, tailoring the experience to their interests and needs. Additionally, be flexible and ready to adjust your approach if the icebreaker isn't going as planned and engaging participants as effectively as you'd hoped. Because the most important part of the activity is to *get* and *keep* people engaged, ensure that the activity remains lively and relevant to everyone involved.

PRO TIP

Use friendly competition to challenge the teams to exceed their goals. For instance, share with them the existing record for the number of innovative ideas generated during similar kaizen events, and then set an ambitious challenge for them to exceed this benchmark.

This approach not only injects a sense of excitement and urgency but also encourages participants to push their boundaries and work collaboratively to achieve (and even surpass) the set target. By framing it as a competitive yet supportive challenge, you can foster a dynamic environment that drives higher performance and more creative outcomes.

REFERENCES

Candra, M., Ginting, R., & Huda, L. N. (2023). Implementation of macro ergonomic participatory in physical work environment: A literature review. *Jurnal Sistem Teknik Industri*, *25*(1), 65–73.

Centers for Disease Control and Prevention (CDC), National Institute for Occupational Safety and Health (NIOSH). (2024, February 26). Elements of an ergonomics program. U.S. Department of Health and Human Services. https://www.cdc.gov/niosh/ergonomics/ergo-programs/index.html

Duraiappah, A. K., Roddy, P., & Parry, J. E. (2005). *Have participatory approaches increased capabilities?* International Institute for Sustainable Development.

Feldman, K. (2023, August 26). *The concept of Kaizen: The key to continuous improvement.* iSixSigma. https://www.isixsigma.com/dictionary/kaizen/

Government of Canada. (n.d.). Process for implementing ergonomics regulatory requirements. Employment and Social Development Canada. Retrieved November 24, 2024, from https://www.canada.ca/en/employment-social-development/services/health-safety/reports/ergonomics.html

Imada, A. S. (2011). Participatory ergonomics: Past, present, and future. *Journal of Human Ergology*, *40*(1–2), 85–89.

Occupational Safety and Health Administration (OSHA). (n.d.). Ergonomics program management guidelines for meatpacking plants. U.S. Department of Labor. Retrieved November 24, 2024, https://www.osha.gov/publications/OSHA3123

Sandel, M., & Wiliams, R. (2024). Participatory budgeting: A growing approach for citizen empowerment in marginalized communities and promoting SDGs. *Sanford Journal of Public Policy*. https://journal.sanford.duke.edu/article/participatory-budgeting-a-growing-approach-for-citizen-empowerment-in-marginalized-communities-and-promoting-sdgs/?form=MG0AV3

Terdoo, F. (2024). Assessing the role of participatory planning approach in enhancing informal settlements upgrading in low income regions. *Discover Global Society*, 2(98). https://doi.org/10.1007/s44282-024-00117-w

2 DEI in the Ergonomics Lexicon

I shared earlier how deeply ingrained I feel Diversity, Equity, and Inclusion (DEI) is in the DNA of our field, yet we often fail to effectively highlight or fully recognize this inherent value. We're just not great marketers. As ergonomics practitioners and the profession as a whole, we naturally bring significant value to our organizations through efforts to enhance employee well-being, improve worker safety, and boost productivity and operational efficiency. And we do this by recognizing diversity and designing work and the work environment to be inclusive and equitable. So, why not celebrate this? If anything, we should be shouting this from the mountaintops!

Just as operational efficiency and safety are intrinsically linked to "ergonomics," so should other business strategies, including DEI. Ideally, we shouldn't be thinking about one without the other. A key challenge, however, is that many of us may not fully realize or are able to articulate the breadth of this impact. One powerful way to make this value more visible is by integrating it into our discussions about ergonomics improvements.

It is commonplace that when benefits of our ergonomics efforts are explored, those benefits are tied to metrics and nomenclature around safety (in terms of risks and injuries reduced or eliminated) and operational efficiency (in terms of productivity gains, waste reduction, or quality improvement). Digging further into the lexicon, Gantt and Shorrock (2017), in their chapter discussing human factors (HF) in media from *Human Factors and Ergonomics in Practice: Improving System Performance and Human Well-Being in the Real World,* noted that when the ergonomics/human factors (E/HF) discipline is further broken down into its separate elements, "ergonomics" is associated with physical injury, especially in office work environments (sometimes presented in a negative light, emphasizing discomfort or injury over the broader benefits of enhancing human-system interactions) while "HF" is linked with failure and "human error," especially involving frontline workers. This shows that even in our own profession, we allow ourselves to be pigeonholed.

Our ergonomics efforts do so much more. We are beginning to see more linkage to other benefits such as well-being and total worker health (although the latter is still considered within the realm of "safety"). But what about benefits and metrics associated with Employee Engagement, Talent Attraction/Retention, and Sustainability? In other words, elements of DEI? The kicker is we're already reaping those benefits; we're just not that great, in my opinion, at articulating it so that others naturally

DOI: 10.1201/9781032658988-4

make that link. And I get it. Changing the public's perception of what ergonomics *is* is a challenge. We know how broad the scope is within our profession, yet we're still reeling from having it be cemented as an "adjective," as in "ergonomic chair" or "ergonomic keyboard". And it has been tough enough just to get people to see beyond our connection to said chairs and keyboards or the automatic pronouncement of:

> Oh! You mean like carpal tunnel!

To see the plethora of benefits our ergonomics efforts bring, one only has to look at ergonomics competitions like the noted Applied Ergonomics Conference Ergo Cup Competition. The innovations showcased there often go far beyond just productivity gains. Many of the solutions presented have a profound impact on accessibility. These real-world examples highlight how our solutions not only enhance workplace performance but also create environments where all employees are empowered to thrive. For illustrative examples that demonstrate how ergonomics improvements are inherently tied to DEI see Figures 2.1–2.7. These are just a few examples of how the benefits seen with ergonomics projects also improve equity and inclusion (in terms of safety, efficiency, and potentially increased opportunities). As we collectively do more tooting of our own horn in how we bring value within this space, the list of practical examples will only continue to grow as we share within the ergonomics community. However, to get the juices flowing, here are some other examples, broad as they are, on how DEI benefits could be expressed when solutions are shared with others (the list was curated as part of the Applied Ergonomics Society DEI Sub-committee initiative to expand the incorporation of DEI benefits in the Applied Ergonomics Conference Ergo Cup Competition):

- Larger font included to accommodate workers with visual impairments.
- Utilized tall man lettering to call attention to products with similar names.
- Redesigned clearance to permit wheelchair user access and operation.
- Expanded capability for those with impaired hearing by emphasizing visual indicators.
- Made instructions more graphics-intensive, accommodating those with literacy challenges.
- Included diverter/guide to minimize the need for fine manipulation, including for operators with fine motor challenges.

Instructions/Communications in various languages: By offering information in multiple languages, you help to ensure that workers can access and comprehend essential instructions, safety guidelines, and operational procedures. This approach helps to bridge language barriers and reduces the risk of miscommunication.

Beyond integrating DEI benefits into our discussions and assessments of ergonomics solutions, there is also work we can do to challenge the broader perception of our field. Gantt and Shorrock went on to further highlight how the media often skews public understanding of ergonomics (more specifically, "HFs"). The focus tends to be on human error or negative outcomes, overlooking the field's true value

FIGURE 2.1 Woman using controls to lift and move equipment (source: seventyfourimages/Envato); Expanded capability for a gender-diverse population: This is probably the most common contribution to DEI efforts as many ergonomics projects focus on reducing physical strain or improving posture. By minimizing those factors, the project inherently increases the accessibility of that task that can be supported by data from anthropometric data resources (e.g., Liberty Mutual Manual Materials Handling Table (Snook & Ciriello, 1991)), which quantify how these changes accommodate various physical characteristics.

in improving system performance and human health. While we're seeing progress in how the field is perceived, there is still work to be done to move past these negative associations.

Take, for example, how the term "human error" places undue blame on the individual, when we know that it is often the system design—rather than the human—that leads to these errors. This misinterpretation of the language feeds into the negative connotations associated with our field. Another pet peeve of mine (besides people dropping the "syndrome" in "carpal tunnel syndrome"—that's just a personal thing, and I am working on it) is when "ergonomics" gets tied directly to negative terms, like when we hear about "ergonomic risk factors" or "ergonomic injury."

Now, I get it—most people mean well, and I'm honestly happy to see management even talking about these issues, so I pick my battles. But I do try to be intentional about how I use the terms. Instead of calling something an "ergonomic injury," I'll refer to it as a "musculoskeletal disorder-related injury" or "MSD risk factor." It is a small change, and I am by all means not perfect at it, but by using more specific and accurate language, I believe we can shift the focus away from the negative and toward the positive, helping to better showcase the true value of ergonomics—and, by extension, the work we do.

FIGURE 2.2 Image of a handle with knurling on the surface & a tennis racket with grip tape on handle (source: photobalance/Envato; FabrikaPhoto/Envato); Example of added friction (knurling or grip tape) to tools: This inherently improves the accessibility for those with grip issues, such as an older population or end-user group.

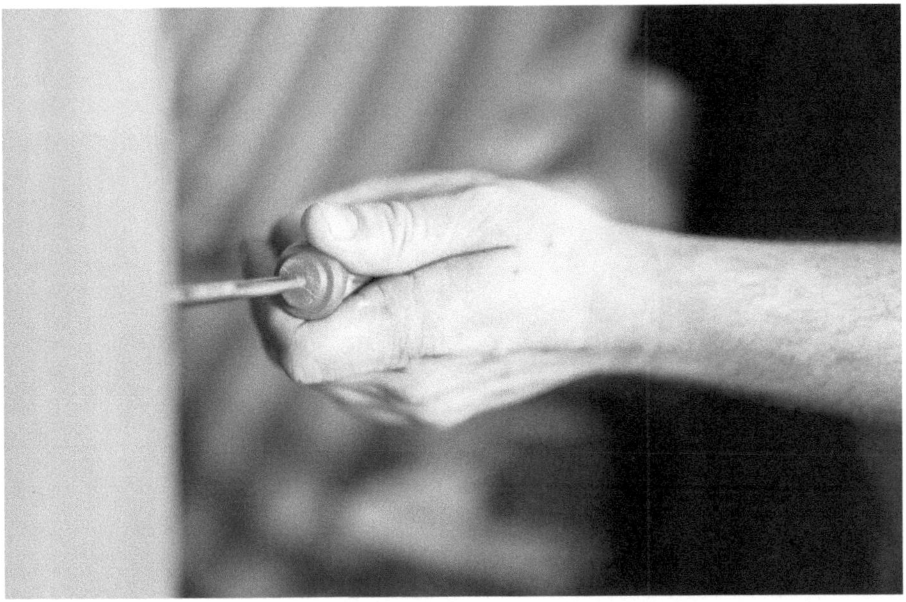

FIGURE 2.3 Image of a person holding a screwdriver (source: daniskim/Envato); Redesigning controls to be used with either hand (or foot): By making controls usable with either hand offers several advantages, particularly in terms of reducing the risk of overuse injuries and accommodating individuals with varying physical abilities such as limited dexterity or strength in one hand/arm/foot/leg.

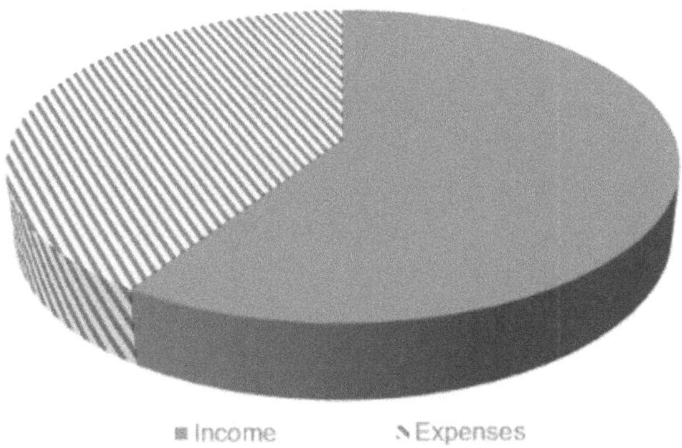

FIGURE 2.4 Image of a pie chart depicting two data points (income and expenses) with different patterns (source: author); Color design consideration for colorblindness: Improvements taking into account the effective use of color (or the inclusion of patterns) to improve the accessibility of information sharing for those with colorblindness—which was the case for me not too long ago, and I'll share that experience (or cautionary tale, rather) later in this book.

FIGURE 2.5 Image of adjustable controls with visual labels (source: DC_Studio/Envato); Inclusive control panel designs: Ensuring that control panels or other interfaces include intuitive designs, such as large, well-spaced buttons with clear labeling or controls with additional feedback, such as tactile or audible feedback. Simplified control interfaces with clear, logical layouts not only improve operational efficiency but it can also benefit users with varying cognitive abilities by reducing the complexity of interactions and potentially reducing cognitive strain.

FIGURE 2.6 Image of a family at home with a man drinking tea while having a video call over a laptop at home (source: drazenphoto/Envato); Flexible work policies that accommodate varied working styles: Alternative working policies, such as remote work or "flex work" time can offer flexibility and opportunities for personalization to support a workforce with diverse behavioral patterns, varied preferred working styles, and varying needs for physical (or social) space and comfort.

FIGURE 2.7 Image of wayfinding airport signs depicting directions (source: enfocaydispara/Envato). Example of making signs/instructions more graphics-intensive to accommodate those with literacy challenges.

REFERENCES

Gantt, R., & Shorrock, S. (2017). Human factors and ergonomics in the media. In S. Shorrock and C. Williams (Eds.), *Human factors and ergonomics in practice: Improving system performance and human well-being in the real world* (pp. 77–91). CRC Press.

Snook, S. H., & Ciriello, V. M. (1991). The design of manual handling tasks: Revised tables of maximum acceptable weights and forces. *Ergonomics*, *34*(9), 1197–1213.

3 DEI in Risk Assessments

I am not going to attempt to write a comprehensive guide on risk assessment tools here. To fully cover all the tools available and provide detailed descriptions for each would require a much longer discussion, and frankly, there are plenty of excellent resources already available that have done this far better than I could. For example, some trusted references for ergonomics and risk assessment tools include:

- American Society of Safety Professionals (ASSP): *How to reduce injuries and illnesses with an ergonomic risk assessment.* (https://www.assp.org)
- American Industrial Hygiene Association (AIHA): *Ergonomics assessment tool kit.* (https://www.aiha.org)
- European Union Occupational Safety and Health (EU-OSHA): *Handbook for ergonomic risk assessments.* (https://osha.europa.eu)
- Applied Ergonomics: *Paper that focuses on assessing ergonomic risks in a range of industries.* (https://www.appliedergonomics.com)

These references can provide more information on the myriads of tools available, when and how they should be used, as well as their advantages and disadvantages. And that's just the beginning—there are many more resources from other countries and organizations that provide valuable insight into ergonomics assessment tools. These references have already compiled detailed explanations, and I highly encourage anyone seeking an in-depth understanding to explore these comprehensive resources. They cover everything from quick, simple assessments to more complex, specialized tools, giving you the necessary tools to conduct thorough and effective risk assessments.

I will, however, discuss the reasons that diversity, equity and inclusion (and DEI as a strategy) are critical elements for consideration in the selection and application of risk analysis tools. For starters, if it hasn't been made clear thus far, by its very nature ergonomics is concerned with considering a wide range of human characteristics, including those related to gender, age, physical abilities, and cultural backgrounds, when designing workplaces and products that fit the people who use them. And it should be clear that humans are not identical, nor are their respective populations. Any ergonomics basic awareness training will highlight some variation of "people come in different shapes and sizes," and thus our goal within ergonomics is to account for that as best we can.

That includes our risk analysis portion of our ergonomics processes. It is just as critical that any ergonomics analysis tool, evaluation, assessment, and consequent intervention are themselves designed to account for the wide range of workplace/end-user diversity. For example, Gurr, Straker, and Moore (1998) recognized the

DOI: 10.1201/9781032658988-5

challenges of transferring risk assessments based on Western environments, such as sitting, when a "*substantial proportion of the world's population is not habituated to chair sitting, but rather to floor sitting and squatting posture*," or the anatomical differences where the distinction between the risk potential would be clear, e.g., the risk associated with extended reaches between individuals with varying functional reaches.

Region specific differences, such as cultural norms and behaviors, are situational-based aspects of our environment, with interaction patterns rooted in beliefs, values, and mental models. For example, differences between cultures with respect to risk perception (Smith-Jackson & Wogalter, 2000), or how language differences can result in varied risk perceptions with hazard ratings (Laughery & Wogalter, 1997). Or how a work or safety culture that perpetuates an exclusive environment, potentially silencing the voices of individuals such as people of color, could negatively impact the risk assessment process by creating discomfort around reporting unsafe conditions (Shore & Chung, 2022).

This means that the tools and evaluation programs used for risk assessments, which will vary in complexity and scope (see aforementioned references), should also account for the varying differences within a diverse population. And we should prioritize using tools that have been tested and proven to work effectively due to consideration given to (or regardless of) age, gender, race, ethnicity, body size, etc.

ANTHROPOMETRY

The body measurements of people can vary significantly, with these measurements being influenced by many factors such as genetics, health conditions, and general lifestyle activities such as exercise levels. As an example, a recent article from Microsoft Network (MSN) (Millhone, 2023) discussed the significant changes in the average male heights across the globe as compared to hundreds of years ago. It was theorized that improved and increased access to nutrition over the years has aided in humans being able to reach their genetic height potential, with, as an example, men from Iran seeing an average height difference increase of 170 mm over the past 100 years, showing how an individual's environment can have an impact on anthropometric measures. These variances are reflected in the anthropometric data utilized to understand and design for the variability within a given population. According to the Institute of Industrial and Systems Engineers (Ed., 1983), anthropometry is the study of people and their physical dimensions. A common tool used among ergonomics practitioners, it constitutes static measurements (measurements of the body at rest, such as stature or eye height, whether sitting or standing) and dynamic measurements (measurements of the body in motion, such as functional arm reaches at various angles). It is with these measurements that we can advise on design parameters for equipment and processes to best fit, in terms of safety and efficiency, the intended end users. Therefore, it is important for us to understand the range and proper use of static and dynamic measurements within diverse populations to help design workspaces that accommodate a diverse workforce.

There is no shortage of available anthropometric datasets. For example, National Institute for Occupational Safety and Health (NIOSH) has publicly available datasets related to specific occupations such as law enforcement (CDC, 2024), firefighters (CDC, 2015) and Emergency Medical Technicians (CDC, 2016), as well as for specific body parts such as the head (CDC, 2016). With regards to general population anthropometric information, the CDC's National Health and Nutrition Examination Survey (NHANES) (CDC, n.d.) program assesses the health and nutritional status of adults and children in the U.S., and that information is utilized to shape public policies. Additionally, there is the U.S. Army Anthropometry Survey (ANSUR II) (The OPEN Design Lab , n.d.), and Civilian American and European Surface Anthropometry Resource (CAESAR) (Society of Automotive Engineers (SAE) International, 2003), derived from both the North American and European populations. In other countries, there is the Health Survey for England (NatCen Social Research, 2024), iMOVE 6-002 Australian Size Variation for Design (Fraysse, 2023) and Germany's MikroZensus (Statistik Bayern, n.d.). Commercially available datasets include PeopleSize from Open Ergonomics (Open Ergonomics, n.d.).

While there is standardization in terms of the collection of anthropometric data and utilization of said data (as outlined in standards such as the International Organization for Standardization (ISO) standard 7250, ISO 15535, and MIL-STD-1472 specific to respective situations), the selection and application of anthropometric data is not without its challenges, as expressed by my colleague Jignesh Vadeher, who, with over 15 years of experience with environmental, health, and safety and Environmental, Social, and Governance in various industries, has a unique perspective working and applying global specifications locally in Asia.

Jignesh recognized that when widely adopted tools and guidelines are used on a global scale, such as the case with many anthropometric data/tools, there can be confusion about whether they are suitable for specific countries or local contexts. He noted from his experience that many guidelines developed in economically advanced countries are frequently copied and pasted into the standards of developing countries without any or very few modifications or adjustments to account for local conditions. Or it may be unclear if they have been tailored to fit the specific needs of individual countries. This can result in underestimating or overestimating local conditions (such as the availability of suggested mitigations), potentially leading to unmet objectives and ineffective solutions.

Additionally, there can be challenges related to the collection of anthropometric data essential for creating accurate population databases. As this data collection involves measuring human body dimensions, there is the potential for the databases to not accurately represent the target population if there is insufficient participation from that group. This lack of representation can lead to significant mismatches, resulting in inadequate risk identification, as highlighted by Alrashdan (2024). Therefore, it is crucial that the datasets applied are relevant and applicable to the populations they are meant to serve.

Given these potential challenges that may hinder our efforts, what steps should we, as ergonomics practitioners, take? Jignesh stressed the importance of first being able to recognize that tools and guidelines shouldn't be applied universally across different countries without considering local contexts. What works well in a developed nation may not translate effectively to a developing country due to variations in social norms, regulatory requirements, and environmental conditions. This view aligns with the arguments presented by Smith-Jackson and Wogalter (2000), who noted that safety designs—and by extension, risk analysis initiatives, including the use of anthropometric tools and guidelines—must be tailored to fit specific cultural contexts. Their research highlights that a one-size-fits-all approach can lead to solutions that are irrelevant and fail to address the unique needs and practices of diverse populations.

As ergonomics practitioners, we need to broaden our skill sets to become more culturally aware to identify potential disconnects. Engaging with local communities and populations representative of our respective workforces is essential to understanding their specific challenges and perspectives. This can be achieved by involving, or ideally embedding, these stakeholders early in and throughout the Participatory Ergonomics process, including during the selection and application of risk analysis tools and guidelines. For example, hosting focus groups or workshops that include local employees from diverse backgrounds, roles, and experiences can facilitate a review of the ergonomics tools used in developing site-specific or organization-specific ergonomics programs. This approach can also enhance education around strategic and tactical ergonomics initiatives, fostering greater buy-in with the site/organization.

Furthermore, enhancing our emotional intelligence is beneficial as we actively seek ways to be more culturally aware and empathetic toward potential challenges. By practicing active listening and seeking continuous feedback, we can improve our initiatives—using the Participatory Ergonomics framework as a foundation—and explore opportunities for additional training in cultural competency.

From an application standpoint, while there has been significant innovation in data collection methods for anthropometric research—marked by advancements in automation, digitization, and the application of sophisticated computational techniques—data for diverse populations is still not always readily available. This lack of data can complicate modeling predictions and make it challenging to ascertain whether datasets are truly representative. Additionally, ensuring the correct population representation across various work situations and environments adds an additional layer of complexity. There have been efforts to bridge the gap in anthropometric data for specific populations within industrial settings. For instance, Castellucci et al. (2020) conducted a study to establish working height, depth, and ideal manual handling heights tailored specifically for Chilean workers, addressing a previously unmet need. Filho et al. (2023) similarly focused on developing data relevant to the Portuguese working population. From a product development perspective, Fu and Luximon (2020) reviewed ear anthropometry and its industrial design applications, highlighting the necessity for universal anthropometric databases that encompass various ethnicities and body parts.

Recognizing the need for more representative and up-to-date databases, ergonomics practitioners should thoroughly investigate the origins of the tools they intend to use, ensuring the datasets are applicable to their specific populations. This requires the user to understand the selected tool's capabilities and limitations, including detailed information regarding the tools' evaluation methods. However, especially in the case of digital human modeling, the underlying algorithms or assumptions that drive the functionalities of the models might not be easily transparent, which can lead to users not fully understanding how the tool accommodates different human characteristics (Ziolek & Nebel, 2003). However, these are still conversations that we, as a profession, should be having with manufacturers of these tools.

From a research perspective, it is crucial to advocate for the inclusion of diverse population representation and real-world work demands in research projects. For example, the University of Minnesota partnered with Toyota on the UMN X Toyota Dynamic Hand Anthropometry Database (University of Minnesota College of Design, n.d.), creating a multifunctional database of 3D hand scans from industrial workers to enhance workspace design for their diverse populations. Additionally, Human Solutions of North America, Inc. initiated the Size NorthAmerica project (Kuebler et al., 2019), collaborating with industry partners and universities to collect and update technical ergonomics and apparel design data (using 3D scanning) to reflect North American ethnic diversity and a broader age range (6–75). This effort aims to provide more relevant industry data that incorporates the anthropometric evolution missing in older datasets.

Ultimately, it is essential for ergonomics practitioners to understand both the work and the population for which it is designed. We must also recognize the potential challenges that could lead to inadequate risk assessments, hindering the goal of inclusion in ergonomics risk management programs. With the aforementioned efforts, ergonomics practitioners can create initiatives and solutions that better align with the cultural values and practices of the intended workforce. This also involves educating others involved in the design process, such as engineers and equipment designers, about these challenges and working together with them as stakeholders to choose the most suitable design tools, ensuring that our diverse workforce is properly evaluated and accommodated.

USING MULTICULTURAL ASSESSMENTS TO ENHANCE OUR ABILITY TO SELECT TOOLS

As noted before, the intent of this chapter is not to provide an exhaustive list of the various qualitative and quantitative risk assessment tools available to us as ergonomics practitioners. And hopefully the references shared earlier provided more insight if that is needed. Regardless of the tools you select and utilize, consideration should be given not only to the typical tool selection criteria—applicability to the job/task in question, ability to identify (and to what extent) the risk factors present, comprehensiveness, practicality (which is a factor of the skill level of the individual(s) conducting the analysis)—consideration should also be given to the applicability of the tools to ensure the inclusion of a diverse workplace.

As with all discussions around diversity, this is a complex topic. And the goal of this book is not to provide a comprehensive or rubric available to cover every possible situation an ergonomics practitioner could encounter. The goal of this book is to provide context, guidance, and resources so that the best decision can be made for the situation.

One such resource is the Association for Assessment and Research in Counseling (AARC) Standards for Multicultural Assessments (AARC, 2018). According to the AARC, the purpose of the standards is to promote best practices in assessment, research, and evaluation in counseling. While counseling is not exactly ergonomics (well, it can feel that way on *many* days), there are still lessons that can be learned and applied to our field. For example, the Standards for Multicultural Assessments are very applicable to considerations we should have when selecting and/or developing processes to select risk assessment tools. As ergonomics practitioners, it could be advantageous to us, in our goal of being more culturally aware, to orient ourselves to these standards to help ensure we are adequately taking the diversity of the populations and environments we seek to assess into account. Of the nine standards provided, these few particularly stand out in reference to our charge:

- *Recognize the impact of subtle biases on multicultural populations and the influence of these biases on the assessment process*—with this standard, we as ergonomics practitioners should be intentional in assuring that any analysis takes into account the diverse body types and physical needs of a multicultural workforce. One quick and very valid example is to include varied body types into the standard risk assessment procedures and not rely only on an "average" worker to stand in for the assessment. This is reflective of the design principle (Z94.2) of designing for most. In design, the typical reference of the 5th percentile female and the 95th percentile male is used as reference to capture, theoretically, 90% of the desired user population. Using this concept, risk analysis similarly should capture the risk potential for these references to better take into account diverse body types (respective to the risk being assessed—e.g., when assessing excessive reaches for a workstation, the risk should be assessed for functional forward reach for the 5th percentile female, or when assessing trunk flexion, the risk should be assessed for the knuckle height of the 95th percentile male). Many tools, such as digital modeling tools, allow for the selection of specific population percentiles for anthropometric models.
- *Identify the impact of intersecting privileged and marginalized statuses on themselves and their clients in regard to the assessment process.* Intersectionality is when different aspects of a person's identity intersect to create unique experiences which can influence the experiences and outcomes of the ergonomics assessment. For example, marginalized individuals may experience heightened stress levels due to discrimination or microaggressions in the workplace, which can exacerbate musculoskeletal disorders (MSDs) or increase the risk for MSDs (Spainierman et al., 2021; Yang et al., 2023). Understanding all the possible challenges from the myriads of intersectionality situations is a significant, if not impossible challenge—and that is not the ask for ergonomics practitioners. The ask is more about having an intersectional lens when conducting assessments or establishing an assessments program. This can look like:

- Learning about intersectionality—there are many training opportunities available to learn more about intersectionality, such as the Intersectionality Resource Guide and Toolkit from the United Nations (UN, 2022).
- Examining your own biases can be especially relevant when designing without consideration for persons with disabilities or who are neurodivergent (e.g., when designing the layout for a new office workspace, failing to account for the diverse preferences of neurodivergent individuals—such as the need for quiet spaces or effective noise reduction—can lead to an environment that does not support their work styles effectively).
- Collectively, with your ergonomics team, thinking through potential scenarios where various intersectionality factors could negatively impact safety and productivity (e.g., discussing how certain forms of safety and ergonomics communications can impact neurodivergent employees, such as those with dyslexia).
- Engage with affected communities—an easy way to do this is to ensure a diverse population is included in risk assessments as well as the development, implementation, and validation of ergonomics mitigations.
- Be intentional about continuously improving your ergonomics process and programs to keep these factors in mind and develop an inclusive ergonomics approach.

- *Seek opportunities for learning by immersion in the social environments of multicultural populations to gain understanding of clients' worldviews.* One philosophy I hold as an ergonomics practitioner is to touch and feel and sense what is happening (as much as I am safely able to do without being extensively trained on a task I am assessing). It is that immersion that helps me understand the perspective of the worker and the context of the work demands. And it is another reason for having employees lead the ergonomics process, because it is their diverse subject-matter-expert perspective that could identify potential challenges that need to be captured in any risk analysis or mitigation implementation.
- *Create awareness of the need for culturally responsive assessment practices with culturally diverse clients.* This can be captured via our ergonomics training programs. By including points about diversity, equity, and inclusion in ergonomics training, as well as our metrics to measure progress—such as the benefits and challenges (see Part III for more details on suggested DEI metrics)—will help to spread awareness and hopefully elicit impactful responses.
- *Address systemic barriers and consider how these barriers impact the interpretation, communication, and use of assessment results.* This means ensuring we have a diverse collection of voices during risk assessments and solution development, paying particular attention to any potential gaps in tool selection (related to reliability and validity), data collection, and the applicability and accuracy of the results (related to measurement error and scores and norms).

ERGONOMICS AND DEI IN THE DESIGN PHASE

As ergonomics practitioners, we recognize the importance of applying ergonomics principles early in the design phase of new processes, products, etc. Doing so fosters a greater recognition and mitigation of the specific ergonomics needs across various departments, enhancing awareness and prioritizing ergonomics considerations in the initial design phase of processes, equipment, etc. This allows for more effective prevention of risks and hazards: prevention through design (PtD).

Identifying and addressing potential risks early on is widely recognized as the most effective way to mitigate and minimize them. However, this common-sense approach has not always been followed. Several factors contribute to this, such as organizational silos that limit collaboration between departments, causing the design process to focus on narrow departmental goals rather than a holistic approach. Additionally, design engineers may lack the knowledge and expertise needed to identify and address safety and ergonomics concerns (Wulff et al., 1999).

In 2007, NIOSH launched the PtD initiative, citing due to continued significant prevalence of occupational injuries and illnesses there is a

> need a comprehensive approach for addressing worker health and safety issues by eliminating hazards and minimizing risks to workers throughout the life cycle of work premises, tools, equipment, machinery, substances, and work processes, including their construction, manufacture, use, maintenance, and ultimate disposal or re-use
>
> **(NIOSH, 2011).**

This initiative is designed to be a proactive approach to workplace safety that focuses on designing out hazards to minimize or eliminate the risks of serious injury or illness through the anticipation and mitigation of potential hazards in various aspects of work. Beyond the prevention of injuries and illnesses, the approach has been touted as leading to significant cost savings by reducing the need for later safety interventions. A collaborative effort through various partners, including industry, labor, trade associations, professional organizations, and academia, the initiative specifically focuses on research into the effectiveness of interventions and exploring new solutions for design-related challenges, education of others to incorporate PtD principles in the design and redesign of workplaces, best practice sharing of effective solutions (to encourage wider adoption), and policy making to integrate PtD into organizations' design standards and guidelines. The ANSI/ASSP Z590.3 standard (in alignment with ISO 31000) provides more regulatory guidance in the implementation of the PtD initiative, including the structure of design reviews, stakeholder roles and responsibilities, and the management of change processes. The latest revision (as of 2024) is ANSI/ASSP Z590.3-2021, and it includes specific guidance related to incorporating ergonomics and human factors engineering principles.

Participatory engagement of stakeholders is a critical element of PtD, with oftentimes the ergonomics expert/practitioner having to serve as the "voice" of the operators to ensure needs, whether physical, cognitive, or psychosocial, are adequately recognized and considered (as will be discussed in Part II). As Wulff et al. highlighted, without participatory and collaborative engagement across all stakeholders it is often the operators, who are directly affected by poor design, who often lack the ability to voice their concerns during the design phase (1999). However, the emphasis of PtD approaches, whether via specific implementation of NIOSH's PtD or the adherence to the ANSI/ASSP or ISO risk management standards, still remains an opportunity to go beyond simply eliminating and minimizing hazards, which, by default, should include considerations for the various needs and barriers of a diverse workforce. The concept of inclusive design seeks to mitigate social inequalities by promoting designs that enable rather than disable. And as Lamirande notes, despite the call and advocacy for inclusive design, many still neglect user perspectives early in the design process, possibly due to fragmented approaches to inclusive design—such as "Universal Design" and "Inclusive Design" principles—leading to confusion and a lack of clear guidance for practitioners (Lamirande, 2022).

Both Universal Design and Inclusive Design are approaches that consist of a set of principles aimed at proactively enhancing the usability and accessibility of systems, including products, processes, and services. These principles guide designers in creating solutions that cater to a diverse range of users, ensuring that people of varying abilities, backgrounds, and preferences can effectively engage with and benefit from the designs.

UNIVERSAL DESIGN

One way to make our world more accessible and inclusive is to apply the principles of Universal Design. This means designing products and environments that can be used by everyone, regardless of their abilities, without requiring any modifications or special features. It prioritizes the development of inherently accessible solutions, aiming for usability across the broadest audience and striving to eliminate barriers from the beginning. The idea of Universal Design really speaks to our goal as ergonomics practitioners. And it seems like common sense: Design for the greatest number of people.

The Centre for Excellence in Universal Design (Centre for Excellence in Universal Design, n.d.) describes it as the creation of environments that can be accessed, understood, and utilized by all individuals, regardless of their age, size, ability, or disability. In simpler terms, it is about good design—design that embraces an inclusive approach by considering diverse needs throughout the process. These needs go beyond just physical impairments; they include any barriers that might hinder a person's engagement with the design. The idea is that when designs are made accessible for those who require them, everyone can benefit, regardless of whether they have specific needs. According to a 2003 Forrester report commissioned by Microsoft, 60% of the adult workforce is likely to gain from using accessible technology, highlighting that creating accessible information and communications technology products and services is crucial not just for individuals with disabilities but for a broader audience (Persson et al., 2015).

The term "Universal Design" was introduced by Ronald Mace, an architect and advocate for inclusive design who was influenced by the challenges he encountered due to his mobility impairment from polio, which required him to use a wheelchair.

Unlike accommodations, which involve modifying or retrofitting a design to meet specific individual needs, Universal Design seeks to eliminate the necessity for such adaptations altogether. This goal is achieved by applying the 7 Principles of Universal Design, developed at North Carolina State University by Mace along with a group of product designers, engineers, environmental researchers, and architects. These principles are intended to guide designers toward more inclusive solutions. The principles, sourced from NC State University's College of Design (NC State Univeristy College of Design, n.d.), include:

Principle 1: Equitable Use—The design is useful and marketable to people with diverse abilities.

- 1a. Provide the same means of use for all users: identical whenever possible; equivalent when not.
- 1b. Avoid segregating or stigmatizing any users.
- 1c. Provisions for privacy, security, and safety should be equally available to all users.
- 1d. Make the design appealing to all users.

Principle 2: Flexibility in Use—The design accommodates a wide range of individual preferences and abilities.

- 2a. Provide choice in methods of use.
- 2b. Accommodate right- or left-handed access and use.
- 2c. Facilitate the user's accuracy and precision.
- 2d. Provide adaptability to the user's pace.

Principle 3: Simple and Intuitive Use—Use of the design is easy to understand, regardless of the user's experience, knowledge, language skills, or current concentration level.

- 3a. Eliminate unnecessary complexity.
- 3b. Be consistent with user expectations and intuition.
- 3c. Accommodate a wide range of literacy and language skills.
- 3d. Arrange information consistent with its importance.
- 3e. Provide effective prompting and feedback during and after task completion.

Principle 4: Perceptible Information—The design communicates necessary information effectively to the user, regardless of ambient conditions or the user's sensory abilities.

- 4a. Use different modes (pictorial, verbal, tactile) for redundant presentation of essential information.
- 4b. Provide adequate contrast between essential information and its surroundings.
- 4c. Maximize "legibility" of essential information.
- 4d. Differentiate elements in ways that can be described (i.e., make it easy to give instructions or directions).
- 4e. Provide compatibility with a variety of techniques or devices used by people with sensory limitations.

Principle 5: Tolerance for Error—The design minimizes hazards and the adverse consequences of accidental or unintended actions.

- 5a. Arrange elements to minimize hazards and errors: Most used elements, most accessible; hazardous elements eliminated, isolated, or shielded.
- 5b. Provide warnings of hazards and errors.
- 5c. Provide fail-safe features.
- 5d. Discourage unconscious action in tasks that require vigilance.

Principle 6: Low Physical Effort—The design can be used efficiently and comfortably and with a minimum of fatigue.

- 6a. Allow users to maintain a neutral body position.
- 6b. Use reasonable operating forces.
- 6c. Minimize repetitive actions.
- 6d. Minimize sustained physical effort.

Principle 7: Size and Space for Approach and Use—Appropriate size and space are provided for approach, reach, manipulation, and use regardless of the user's body size, posture, or mobility.

- 7a. Provide a clear line of sight to important elements for any seated or standing user.
- 7b. Make reach to all components comfortable for any seated or standing user.
- 7c. Accommodate variations in hand and grip size.
- 7d. Provide adequate space for the use of assistive devices or personal assistance.

INCLUSIVE DESIGN

Inclusive design goes beyond accessibility and usability to consider the social and emotional aspects of human diversity. This approach emphasizes engaging with specific users or user groups to understand their unique challenges, aiming to remove barriers through accommodatable design. As with Universal Design, there are principles widely accepted as guides for Inclusive Design. As defined by the U.K's Design Council (Design Council, n.d.) they include:

- Inclusive: So everyone can use them safely, easily, and with dignity.
- Responsive: Taking into account of what people say they need and want.
- Flexible: So different people can use them in different ways.
- Convenient: So everyone can use them without too much effort or separation.
- Accommodating: For all people, regardless of their age, gender, mobility, ethnicity, or circumstances.
- Welcoming: With no disabling barriers that might exclude some people.
- Realistic: Offering more than one solution to help balance everyone's needs and recognizing that one solution may not work for all.

UNIVERSAL DESIGN VS. INCLUSIVE DESIGN

If you're anything like me, you might have used "Universal Design" and "Inclusive Design" interchangeably. However, I've come to realize there are subtle yet important differences between the two, and understanding these distinctions has helped me appreciate their respective roles in our goal of "designing for all."

While Universal Design refers to concepts that are applicable to everyone, Inclusive Design takes a more personalized approach, focusing on the specific needs—whether physical, cognitive, or emotional—of individual users or groups. To illustrate this difference, consider how *I* distinguish the two using an analogy between User Interface (UI) design and User Experience (UX) design. In my view, good UI design, akin to Universal Design, ensures that everyone can access and use an app, regardless of their abilities or backgrounds. On the other hand, good UX design, which to me aligns with Inclusive Design, ensures that everyone feels valued and has an enjoyable experience while using the app. Or as my colleague Richard Goggins so aptly said, "Design for populations, customize for individuals."

The concept of designing for everyone via whatever approach does not come without its challenges. For one, you can probably see how the terms can become conflated with each other or with other terms such as "accessibility design" or "design for all." These terms tend to be linked and used more within certain professions and communities. And while different in terms of interpretations and methodologies, they strive for the same goal, which is removing barriers to increase accessibility. However, these differences and the general lack of consensus for the definition of "accessibility" (and its general ambiguity) have led to confusion and disagreements within the overall design community and possibly the reasoning for limited adoption of accessibility efforts through standards and guidelines (Persson et al., 2015).

Another challenge with the implementation of Universal Design or Inclusive Design principles would be the lack of known standardized solutions for designs, given that design contexts and user needs often vary widely. This can lead to uncertainty for designers on how to develop spaces or products that are truly accessible and inclusive. I recognize that it is a challenge for us as well as ergonomics practitioners and the profession as a whole, but it is a challenge we are well equipped to take on, hence this book!

Many people are simply unaware of the design concepts and may not know how to acquire the skills necessary to implement these principles. This lack of awareness can impede the development of genuinely accessible and inclusive designs. However, education and collaboration represent effective ways to address these challenges, presenting an opportunity for us as ergonomics practitioners. By informing designers, engineers, and other stakeholders about these concepts and collaborating with them, we can create solutions that cater to the diverse needs of all users while also enhancing understanding of how these design principles align with and benefit from ergonomics. The ergo Trojan horse, if you will.

Regardless of the approaches taken, Lamirande recommends that despite the complexities of focusing on being inclusive within the design phase, such as a lack of a champion to push for inclusion, unconscious bias on the part of the design team, or the lack of effective recruitment approaches to get marginalized individuals to participate, there is a necessity to include marginalized voices in design processes to foster truly inclusive outcomes.

Another key insight from Lamirande's research, which involved extensive discussions with design practitioners, is the value of viewing inclusive design as an "attitude" or "mindset" rather than a definitive goal to achieve. This perspective can help mitigate the challenges arising from the absence of clear guidance. Given the complexity of inclusive needs—both in relation to a diverse workforce and the constantly evolving work environment—it is unrealistic to expect a comprehensive checklist that addresses every possible requirement. Like with NIOSH's PtD (CDC, n.d.), Universal Design, and Inclusive Design, success relies on the knowledge, application, and advocacy of these principles. This, too, presents a unique opportunity for ergonomics practitioners to foster and solidify this inclusive mindset within our organizations.

REFERENCES

Alrashdan, A. (2024). Occupational safety and health for adult Saudi Arabian women: Utilizing national anthropometric data. *Healthcare, 12*(1), 109. https://doi.org/10.3390/healthcare12010109

American Industrial Hygiene Association (AIHA). (n.d.). Ergonomics assessment tool kit. https://www.aiha.org

American Society of Safety Professionals (ASSP). (2023). How to reduce injuries and illnesses with an ergonomic risk assessment. https://www.assp.org

Applied Ergonomics: Tools and Methods for Improving the Fit of the Workplace. (n.d.). Paper that focuses on assessing ergonomic risks in a range of industries. https://www.appliedergonomics.com

Association for Assessment and Research in Counseling (AARC) Standards for Multicultural Assessments. (2018). AARC standards for multicultural assessments. https://aarc-counseling.org/wp-content/uploads/2020/04/AARC-Standards-for-Multicultural-Assessments-2018.pdf

Castellucci, H., Viviani, C., Arezes, P., Molenbroek, J. F., Martínez, M., Aparici, V., & Dianat, I. (2020). Applied anthropometry for common industrial settings design: Working and ideal manual handling heights. *International Journal of Industrial Ergonomics, 78*, 102963. https://doi.org/10.1016/j.ergon.2020.102963

Center for Universal Design. (n.d.). *Center for universal design*. North Carolina State University. Retrieved August 9, 2024, from https://design.ncsu.edu/research/center-for-universal-design/

Centers for Disease Control and Prevention (CDC), National Center for Health Statistics (NCHS). (n.d.). *About NHANES. U.S.* Department of Health and Human Services. https://www.cdc.gov/nchs/nhanes/about_nhanes.htm

Centers for Disease Control and Prevention (CDC), National Institute for Occupational Safety and Health (NIOSH). (2015). *RD-1007-2015-0 dataset*. U.S. Department of Health and Human Services. https://www.cdc.gov/niosh/data/datasets/RD-1007-2015-0/

Centers for Disease Control and Prevention (CDC), National Institute for Occupational Safety and Health (NIOSH). (2016). *RD-1008-2016-0 dataset*. U.S. Department of Health and Human Services. https://www.cdc.gov/niosh/data/datasets/rd-1008-2016-0/default.html

Centers for Disease Control and Prevention (CDC), National Institute for Occupational Safety and Health (NIOSH). (2024). *RD-1081-2024-0 dataset.* U.S. Department of Health and Human Services. https://www.cdc.gov/niosh/data/datasets/rd-1081-2024-0/default.html

Centers for Disease Control and Prevention (CDC). (n.d.). *Prevention through design: Overview.* National Institute for Occupational Safety and Health (NIOSH). Retrieved October 18, 2024, from https://www.cdc.gov/niosh/ptd/about/index.html#:~:text=Prevention%20through%20Design%201%20Overview%20One%20of%20the,3%20Approach%20...%204%20News%20%26%20Events

Centre for Excellence in Universal Design. (n.d.). *Definition and overview.* Centre for Excellence in Universal Design. Retrieved April 22, 2024, from https://universaldesign.ie/about-universal-design/definition-and-overview

Design Council. (n.d.). *The principles of inclusive design.* Design Council. Retrieved August 9, 2024, from https://www.designcouncil.org.uk/fileadmin/uploads/dc/Documents/the-principles-of-inclusive-design.pdf

European Union Occupational Safety and Health (EU-OSHA). (n.d.). Handbook for ergonomic risk assessments. https://osha.europa.eu

Filho, P. C. Anacleto, da Silva, L., Mattos, D., Pombeiro, A., Castellucci, H. I., Colim, A., Carneiro, P., & Arezes, P. (2023). Establishing an anthropometric database: A case for the Portuguese working population. *International Journal of Industrial Ergonomics, 97,* 103473. https://doi.org/10.1016/j.ergon.2023.103473

Fraysse, F., Wade, A., Furnell, A., Kirsch, C., & Murray, P. (2023). *iMOVE 6-002 Australian size variation for design M004: Detailed anthropometry dataset.* University of South Australia.

Fu, F., & Luximon, Y. (2020). A systematic review on ear anthropometry and its industrial design applications. *Human Factors and Ergonomics in Manufacturing & Service Industries, 30*(3), 176–194. https://doi.org/10.1002/hfm.20845

Gurr, K., Straker, L., & Moore, P. (1998). Cultural hazards in the transfer of ergonomics technology. *International Journal of Industrial Ergonomics, 22*(4–5), 397–404. https://doi.org/10.1016/S0169-8141(97)00094-2

Institute of Industrial Engineers. (Ed.). (1983). *Industrial engineering terminology: A revision, consolidation, and redesignation of ANSI Z94 index and ANSI Z94. 1-12.* Wiley-Interscience.

Kuebler, T., Luebke, A., Campbell, J., & Guenzel, T. (2019). Size North America – The new North American anthropometric survey. In V. Duffy (Ed.), *Digital human modeling and applications in health, safety, ergonomics and risk management. Human body and motion* (Lecture Notes in Computer Science, Vol. 11581, pp. 85–94). Springer. https://doi.org/10.1007/978-3-030-22216-1_7

Lamirande, M. (2022). Aspects of designing inclusively from practitioner perspectives. *Architecture, 2*(3), 497–517. https://doi.org/10.3390/architecture2030028

Laughery, K. R., & Wogalter, M. S. (1997). Warnings and risk perception. *Handbook of Human Factors and Ergonomics, 2,* 1174–1197.

Millhone, C. (2023). *What is the average male height?* MSN. https://www.msn.com/en-us/health/medical/what-is-the-average-height-for-men/ar-AA1c5DgA?ocid=winp2fptaskbarhoverent&cvid=c1f0fb2ff82d420da74da932943c7810&ei=15

NatCen Social Research, University College London, Department of Epidemiology and Public Health. (2024). *Health-survey for England* [data series]. 8th Release. UK Data Service. SN: 2000021. http://doi.org/10.5255/UKDA-Series-2000021

National Institute for Occupational Safety and Health (NIOSH). (2011). *Ergonomics and the prevention of musculoskeletal disorders: A guide for the workplace.* https://doi.org/10.26616/NIOSHPUB2011121

NC State University College of Design. (n.d.). Center for Universal Design. https://design.ncsu.edu/research/center-for-universal-design/

Open Ergonomics. (n.d.). PSZ. Retrieved August 3, 2024, from https://www.openerg.com/psz/

Persson, H., Åhman, H., Yngling, A. A., et al. (2015). Universal design, inclusive design, accessible design, design for all: Different concepts—one goal? On the concept of accessibility—historical, methodological, and philosophical aspects. *Universal Access in the Information Society, 14*(4), 505–526. https://doi.org/10.1007/s10209-014-0358-z

Shore, L. M., & Chung, B. G. (2022). Inclusive leadership: How leaders sustain or discourage work group inclusion. *Group & Organization Management, 47*(4), 723–754.

Smith-Jackson, T. L., & Wogalter, M. S. (2000). Applying cultural ergonomics/human factors to safety information research. *Proceedings of the IEA 2000/HFES 2000 Congress.* https://www.safetyhumanfactors.org/wp-content/uploads/2020/07/191%29SmithJackson,Wogalter%282000%29%28a%29.pdf

Society of Automotive Engineers (SAE) International. (2003). *CAESAR® final report (RR-017).* SAE International.

Spanierman, L. B., Clark, D. A., & Kim, Y. (2021). Reviewing racial microaggressions research: Documenting targets' experiences, harmful sequelae, and resistance strategies. *Perspectives on Psychological Science, 16*(5), 1037–1059. https://doi.org/10.1177/1745691621101994

Statistik Bayern. (n.d.). Mikrozensus: Bevölkerungsstatistik. Retrieved August 3, 2024, from https://www.statistik.bayern.de/statistik/gebiet_bevoelkerung/mikrozensus/index.html

The OPEN Design Lab. (n.d.). ANSUR II. The Pennsylvania State University. https://www.openlab.psu.edu/ansur2/

UN Women. (2022). *Intersectionality resource guide and toolkit.* United Nations Women. https://www.unwomen.org/sites/default/files/2022-01/Intersectionality-resource-guide-and-toolkit-en.pdf

University of Minnesota College of Design. (n.d.). UMN x Toyota dynamic [Research project]. University of Minnesota. https://design.umn.edu/research/research-creative-scholarship-engagement/research-projectsHere are the remaining APA citations formatted:

Wulff, I. A., Westgaard, R. H., & Rasmussen, B. (1999). Ergonomic criteria in large-scale engineering design—II Evaluating and applying requirements in the real world of design. *Applied Ergonomics, 30*(3), 207–221. https://doi.org/10.1016/S0003-6870(98)00030-1

Yang, H., Lu, M.-L., Haldeman, S., & Swanson, N. (2023). Psychosocial risk factors for low back pain in US workers: Data from the 2002–2018 quality of work life survey. *American Journal of Industrial Medicine, 66*(1), 41–53. https://doi.org/10.1002/ajim.23458

Ziolek, S., & Nebel, K. (2003). Human modeling: Controlling misuse and misinterpretation. *SAE Technical Paper 2003-01-2178.* https://doi.org/10.4271/2003-01-2178

4 DEI and Solution Development

How many times have you hit the wall, so to speak, trying to find a solution to a particularly challenging problem? If you haven't faced this yet, rest assured, you will at some point. If you have experienced it, you probably know the frustration that comes with it, along with a sense of inadequacy or self-doubt. It is a common feeling that can be both paralyzing and disheartening. But can I tell you a secret? There *is* a solution—it is called sharing the load.

The old saying, "a problem shared is a problem halved," holds a great deal of truth. By reaching out and involving others in solving a problem, you're not just looking for one solution—you're opening the door to a variety of potential solutions. Engaging others brings diverse perspectives shaped by their unique experiences and expertise, allowing the problem to be viewed, assessed, and addressed in different ways. This collective input can spark new approaches and creative ideas that you might not have thought of on your own, while also enhancing the problem-solving skills of all involved. This collaboration fosters a sense of ownership and pride, ultimately leading to more sustainable outcomes.

This is a lesson I learned early in my career when my first experience with ergonomics began with an employee-driven Participatory Ergonomics process—a lesson I'm truly grateful to have learned.

FINDING SOLUTIONS

Developing effective solutions for ergonomics-related challenges is a key component of a successful ergonomics process. However, it can often be very challenging to find the right solution, especially when traditional fixes or well-known solutions don't apply. Sometimes, what is needed is a completely new and innovative solution or approach. In these situations, it is essential to employ diverse strategies that boost creativity and inspire fresh ideas, both for us and our teams. Using varied problem-solving techniques and leveraging different perspectives can inspire innovative solutions that address the unique challenges at hand and drive the success of our ergonomics initiatives.

Tapping into the Creativity of the Workforce

What do Post-It® Notes, zip codes, and the birth of "software engineering" have in common? They were all brainchildren of individuals who were not what you would consider your "traditional" innovator—meaning, they didn't have "INNOVATOR" in their job title.

 DOI: 10.1201/9781032658988-6

Post-It® Notes were invented by Spencer Silver and Arthur Fry, employees of 3M, who stumbled upon the invention when a low-tack adhesive was developed by accident (3M, n.d.). The idea of zip codes, which seems like a no-brainer now, was invented by Robert Moon and H. Bentley Hahn, postal workers for the United States Postal Service (History.com Editors, 2020). They devised a numbering system to divide the country into geographical areas, enhancing sorting accuracy and efficiency. Lastly, Margaret Hamilton, a software engineer, led the development of the onboard flight software for the Apollo missions. She coined the term "software engineering" and wrote the code that enabled astronauts to land safely on the moon (Meyer, 2019).

These examples show that innovation can come from anyone, regardless of their job title or background. It is a reminder that great ideas can emerge from unexpected places and people. And while I could go on and on about interesting inventions that changed the world, the point is to share how many of these inventions were not the product of a dedicated group of people solely tasked with developing improvements for their industries. They were invented by workers who saw opportunities based on their individual experiences and perspectives. It was the culture of tapping into the diverse ingenuity of those who may be exposed to challenges in different ways that led to great breakthroughs. And as ergonomics practitioners, we get the opportunity to provide support systems that can help those ideas (and workers) flourish!

As ergonomics practitioners, we are often at the forefront of facilitating that type of thinking from the subject matter experts within our organizations—the employees. In Chapter 1, I spoke about how diversity can and should be leveraged to build effective ergonomics teams and processes. Here in this chapter, I will expound on how it is only natural that we have value to add to any company's innovation strategy.

To remain competitive, adapt to change, and drive sustainable growth, many companies are recognizing the talent within their organizations and taking steps to harness it by promoting a culture of "intrapreneurship." While it may sound like a buzzword, being a blend of "intra" and "entrepreneurship," it represents a genuine movement that is having a positive impact on innovation (Lechner & Gudmundsson, 2014) and employee engagement (Urban & Wood, 2015). This trend is gaining momentum and is likely to continue, extending beyond large corporations to include the public sector, third-sector organizations, and small to medium-sized enterprises (Attle, 2024).

But what exactly is intrapreneurship?

Intrapreneurship is when established corporations foster entrepreneurial mindsets in their employees. This can vary from how the culture of internal innovation is supported—such as through creative campaigns or structured organizational departments—but when employees think like entrepreneurs, it can lead to change and value creation that can increase the success of the business (Porumboiu, 2021). Remember the examples I shared earlier? I think we can all agree that they are definitely factoring into the success of their respective organizations! Not only can the ideas benefit the company in terms of financial success through process improvements or new business ventures, but it can also lead to improved employee experience, which can then lead to higher employee retention. A potential win-win for all involved.

Imagine how diversity can significantly contribute to the development of creative and effective ideas, be it a process improvement that has the power to significantly improve process efficiencies or even the next multimillion-dollar business idea for the organization. As discussed throughout this book, a diverse team can greatly enhance solution development by fostering collaboration among individuals with varied perspectives and experiences. This is especially true for a diverse team equipped with the tools and skills that drive intrapreneurship within an organization.

In this context, our ergonomics processes and overall approach can support any company's innovation strategy. One challenge many companies face with intrapreneurship initiatives is the need for a systematic approach to sustain and maintain their effectiveness. Our expertise in prioritizing, validating, and implementing ideas within our ergonomics processes can be highly beneficial in this aspect. We have extensive experience in drawing ideas from a diverse workforce, making our grassroots efforts more effective compared to a top-down approach. Our trial-and-error efforts often follow the "build-test-learn" approach to innovation, and in many cases, we already have a diverse, creative, and enthusiastic group of employee leaders ready to champion innovation. After all, it is what they do best! Clearly, ergonomics deserves a seat at the innovation table!

In addition to enhancing the innovation framework, all ergonomics processes can benefit from it as well. Whether it is by learning from the ideas generated during an innovation sprint or campaign or by collaborating with innovation teams to refine our solution development skills (e.g., coaching our teams on design thinking principles (to be explored next)), there are numerous opportunities to improve our ergonomics processes. Therefore, it is crucial for ergonomics practitioners and teams to collaborate closely with any innovation initiatives, as this partnership will offer valuable support in idea generation and implementation.

Design Thinking

When tasked with developing solutions, it is critical to first accurately define the problem at hand. In the realm of ergonomics, using frameworks like Design Thinking can help guide us to identify many effective solutions for the challenges we face.

Design Thinking is a human-centered systematic approach to innovation that can be highly effective in developing solutions for ergonomics-related challenges. That definition alone should get our ergo ears to perk up! With Design Thinking, we are deploying a method to develop solutions that are user-centered, innovative, and effectively meet the needs of those they are designed for. It emphasizes and encourages creativity and collaboration through a deep understanding of end-user needs achieved through their engagement and participation throughout the process. While there are several global process models to design thinking (Waidelich et al., 2018), the approach focused on within this book relies on the process as described by Stanford University's design school ("d.school"), which consists of five steps (Stanford University Institute of Design, 2016):

EMPATHIZE • DEFINE • IDEATE • PROTOTYPE • TEST

Let's delve into each step to see how specifically they can apply to our ergonomics activities for a diverse workforce.

Empathize

The initial step in our Design Thinking approach involves empathizing deeply with individuals facing challenges. Personally, this is the most important step as it not only sets the stage for the development of solutions to address the true issues but it begins to lay the foundation for a culture that not only hears the voice of the end-users—or the workers in many of our cases—but also engages them. The purpose of this Empathize step is to develop a profound understanding of their needs and the environment they operate within. And when you're working with a diverse working population, their individual experiences can influence their interactions with processes, machinery, tools, etc. Therefore, it is important to include as many diverse voices within this step as possible.

When tackling an ergonomics-related issue—like addressing a high incidence of musculoskeletal disorders (MSDs) in a specific workshop, department, or office—this step necessitates us to actively listen, shadow workers within the affected area, and conduct interviews. Through these interactions, we will be able to uncover the true realities of their work and their work environment. It is crucial that we not solely rely on prescribed procedures; understanding the actual work dynamics is essential. This reminds me of what my ergonomics colleague Didier Lafont in France had to say about ergonomics and the nature of work (translated from French):

> Ergonomics aims to consider human work as dynamic, variable, changing, not stabilized. It is a precarious balance, subject to complex regulation accessible only through detailed, comprehensive analysis of the activity, considering certain violations, migrations in the face of work standards not as an abnormality, but rather as a process. Permanent adaptation.

This means that we, as ergonomics practitioners, need to approach work as a diverse, dynamic, and ever-changing process rather than a stable and static one. We know there can be a considerable gap between what people are assumed or expected to do and what they actually do, even with standardized work. We must acknowledge that work and work environments are subject to complex changes—either deviations from the standard or reactions to abnormalities—thus we must be prepared to conduct detailed, comprehensive analyses of work activities to consider and analyze the deviations and adaptations to work, not as abnormalities, but as ongoing processes of permanent adaptation.

To effectively do this, we must engage directly with workers and empathize with them. This allows us to observe firsthand the realities of how they move and interact with tools or machinery in their tasks. By listening closely and understanding their perspectives, we can set aside any biases we may have and better grasp the root causes of the issues. This approach not only helps us see the realities of work, but it helps us pinpoint areas where MSD risks and workflow disruptions occur.

PRO TIP

If possible and safe to do, experience the task for yourself as the ergonomics practitioner and encourage others, such as the ergonomics or delegated kaizen team. This firsthand experience can provide valuable insights into the challenges and nuances of the task. Engaging directly with the task can help identify potential issues more accurately and foster a deeper understanding of the solutions needed.

Define

With the Define step, our goal is to define the problem statement based on insights gathered during the Empathize step. This is when we delve more into the root causes of the issues by taking information from the observations and interviews to define the core issues affecting users. Techniques such as process mapping, fishbone diagrams, or the five Whys help further pinpoint where solutions are most needed.

- Process Mapping (Figure 4.1): A visual representation or diagram that illustrates the steps, actions, and decisions involved in a process from start to finish; have workers identify steps within the process where risk is present.
- Fishbone Diagrams (Figure 4.2): Cause-and-effect diagrams that are used to visually represent the various factors that contribute to a problem or outcome; divide the fishbones into the five Ms and work through the causes and ultimate root causes for how each could potentially lead or has led to the MSD (Note: Methods and terminology to divide the fishbone may vary):
    Manpower: Human factors such as skills, training, staffing, and workload.
    Method: Processes, procedures, and work methods.

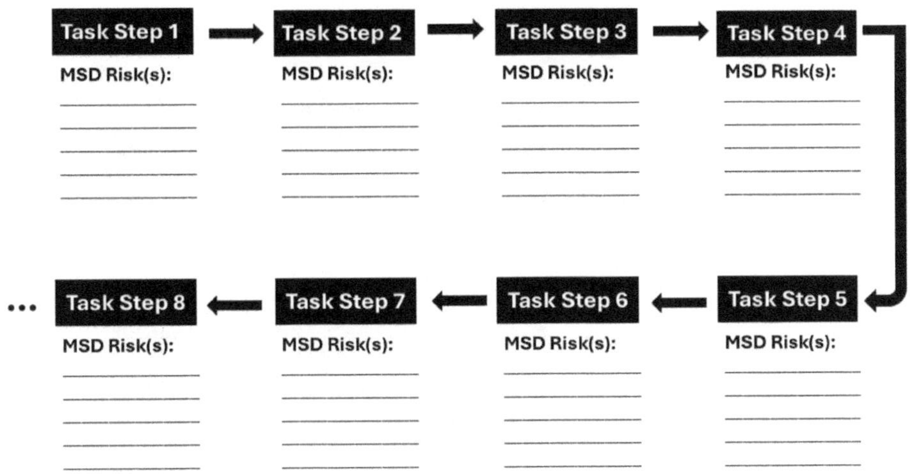

FIGURE 4.1 Image of a simple process map with task steps separated out with area to capture MSD Risk under each task step. (source: author)

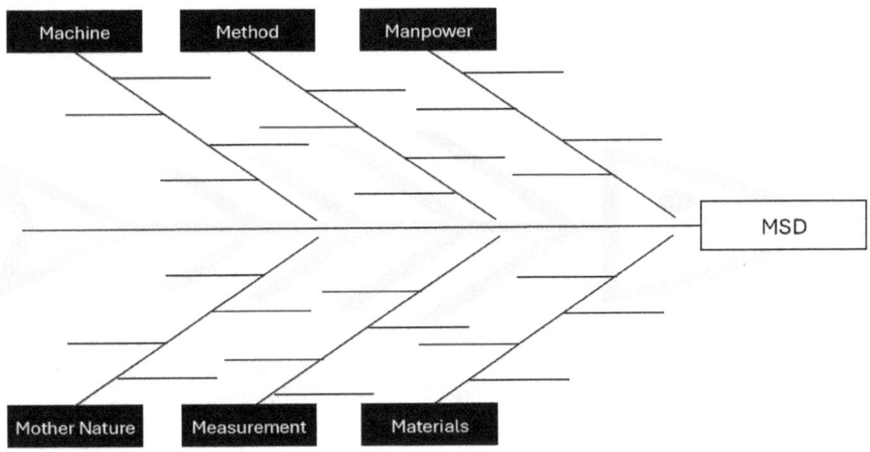

FIGURE 4.2 Image of a fishbone diagram with root cause categories. (source: author)

Machines: Equipment, tools, and technology.
Materials: Raw materials, components, and supplies.
Mother Nature (Environment): Physical work environment, temperature, lighting, noise, vibration, etc.
Measurement: Metrics, data collection methods, and performance indicators or key performance indicators.

- The Five Whys (Figure 4.3): technique where "WHY?" is repeatedly asked until the underlying cause is identified (Note: It is not uncommon for there to be multiple potential root causes related to ergonomics-related risk).

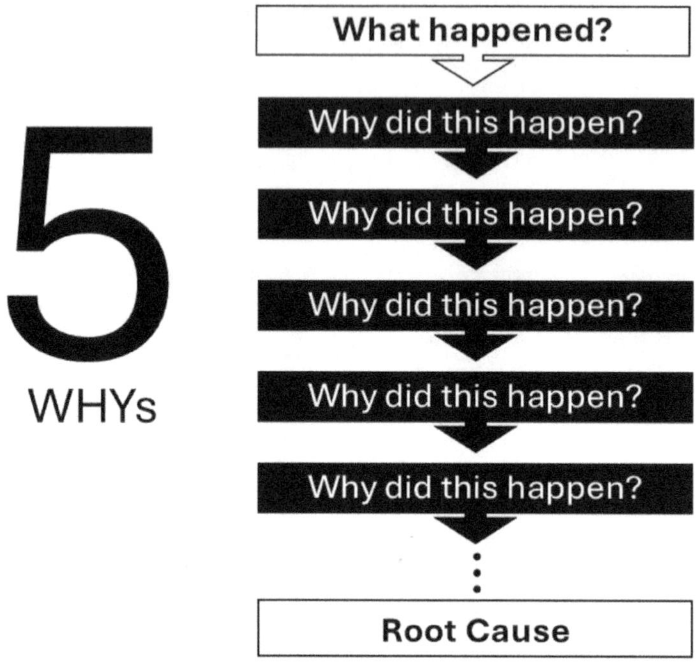

FIGURE 4.3 Image of a process depicting the "Five Whys"; includes a series of "why did this happen?" questions to arrive at root cause. (source: author)

When articulating the problem statement, it is important to frame it correctly so that the issue *as it relates to the end user* is clearly defined, also known as "problem framing." In ergonomics-related examples, the problem statement could be worded as: "Shorter stature workers are having to handle heavy loads above shoulder level during the loading task and are experiencing injuries and discomfort in the shoulders" or "Workers are experiencing MSDs to the back due to heavy lifting at a high pace during the unloading task." Note: It is not uncommon to identify multiple problems, with each requiring its own focus during an ergonomics risk assessment. That is why it is critical to identify, define, and prioritize all relevant problems so that adequate resources can be allocated to address each. Additionally, by intentionally including the experience of the workers in the problem statement, we bring the focus back on the user—the intent of the Design Thinking approach. It is also a reminder to consider the diverse characteristics of the workers impacted when ideating.

Ideate

The next step in the Design Thinking approach is Ideate. This is when a wide range of creative solutions are generated to address the defined problem. The goal here is to develop as many ideas as possible to expand the range of potential solutions.

Eventually, they will be narrowed down, but in the beginning, multiple ideas should be laid out as ideas can be revisited or diverged into new directions at a later time. There are many techniques that can be deployed to gather ideas, but a basic step is to prompt the process with "How Might We..." The question is intended to be open-ended to encourage a range of potential suggestions. This simple reframing from what we traditionally start with (i.e., "What Can Be Done..."), along with a problem statement focused on the needs of the user as opposed to the system turns the problem into a design task with no implied "correct conclusion" as opposed to a problem-solving task with an implied "clear single solution." Different mental strategies are employed with each, with the former associated with divergent thinking considered most advantageous for idea generation (Vries et al., 2019; Daikoku et al., 2021) and activating the right dorsolateral prefrontal cortex of the brain (Gilbert et al., 2010). The inclusion of "We" in the question also invites collaboration to collectively identify diverse ways to address the problem statement, which includes the overall experience of the worker.

With the initial "How Might We" prompt, how ideas are captured can be quick and informal, such as having a group jot down *any* idea that comes to mind within a set timeframe (usually only a few minutes) and then dedicating more time to build off them, or they can be elaborate structured events, such as mind mapping or storyboarding. Regardless of the technique used, the goal should be to develop as many ideas as possible. Quantity is key!

In the Define and Ideate steps, it is crucial to avoid biasing ourselves with preconceived notions about the problem and potential solutions before completing the Empathize step. It is challenging not to enter a situation with preconceived ideas. However, doing so risks falling into a "when all you have is a hammer, everything looks like a nail" mindset. Trying to impose a solution prematurely fails for several reasons: Firstly, it may not address the root cause of the problem; secondly, skipping Empathize risks lacking worker/stakeholder buy-in due to perceived inauthenticity (and trust me, they can tell when you're being inauthentic); and lastly, it hampers the generation of new ideas. That's why, during the ideation phase of an ergonomics solution development kaizen event, I deliberately delay reviewing past solutions. Approaching solutions like a checklist or a shopping cart can significantly stifle creativity, which is vital for fostering innovation within our profession.

Prototype

With this step, the goal is to tangibly develop selected ideas so that more information can be gleaned to further improve upon the idea. A prototype can take many forms: an idea of a layout sketched out on a piece of paper, the shape of a tool shaped with modeling clay (as the smell of your childhood wafts in the air!), crude drawings done in PowerPoint, or the like. But what it should not be is perfect. Trying to create a perfect prototype can have you expending unnecessary time and effort, and it could actually discourage feedback from participants reluctant to share anything constructive, assuming the perceived effort that went into making it perfect. You want the prototype to be a rough representation of what you want your idea to be/do so that

1) useful feedback can be given, 2) the idea can be compared to other ideas, and 3) it can be improved upon.

As ergonomics practitioners, we often deal with non-physical solutions, such as processes, experiences, and techniques. It may be worth noting that these, too, can be prototyped. Consider using storytelling in these cases to get the point across and have participants visualize all the steps to be able to give feedback on where issues could occur or give suggestions for improvements. For example, if the idea of a new process needs to be prototyped, you could begin to craft your story by first introducing a main character. From there, describe that character's daily routine and the challenges they face with the current process. Then describe the character's specific struggles with the process, highlighting the risks, such as strain on their shoulders due to the awkward posture required by the current process, and how this adversely impacts their quality of life. After the problem has been clearly articulated, introduce the process improvement solution and describe how this new process reduces the need for the character to awkwardly reach. Be sure to detail the positive impact of the new process on the character's work experience, explaining how specifically it would enhance their comfort and safety and improve operational efficiencies, as well as any other organizational benefits. Incorporate time for those hearing the story to give feedback on any suggested adjustments, having them consider what their experience may be with the process.

A compelling narrative achieves several crucial objectives—it makes abstract concepts concrete by providing the necessary details to help others vividly imagine not just the current challenges but also potential solutions. More importantly, in my opinion, it humanizes issues that are often obscured by statistics and metrics. In our field, while we monitor data to track progress, our primary focus is on enhancing people's lives—whether through improved safety, operational efficiency, well-being, quality, or employee experience. I've observed that our profession could benefit greatly from effective marketing, and mastering the art of storytelling, especially in personalizing experiences, is a valuable and practical skill we all need to master.

Test

The final step within the Design Thinking approach is Test. As you can imagine, this step works in tandem with the Prototype step, as this is when you evaluate ideas/prototypes with users to gather feedback and insights for further improvement. Typical methods used in this phase include user testing, where end users or workers interact with prototypes and provide valuable feedback, as well as qualitative and/or quantitative data on aspects such as effectiveness in mitigating ergonomics-related risks, usability, ease of use, and overall user experience. Additionally, iterative improvements are made by using insights from testing to refine and adjust the prototypes until a viable solution is achieved.

One important thing to remember about the Design Thinking approach is that while it may be tempting to assume that the steps are linear and that one must happen before the other, the truth is there is no one true way to move through the approach. In fact, you may find yourself going in circles—but in a good way! Because that means you are revisiting steps as you learn more. And that is ideal. We should be taking any new information or lessons learned throughout the process and revisiting steps to ensure we're on the right path. For example, as you begin to develop prototypes of solutions and test them, the feedback you receive from the end users/workers may lead you down the path of a completely different solution. Or you may find that you missed the mark on defining the problem and the better place to focus efforts is another issue later identified throughout the ideate step. We should be comfortable with constantly conducting experiments and letting those new discoveries change our perspectives.

LEVERAGING THE DIVERSITY AROUND US FOR IDEAS

In conjunction with using the Design Thinking approach as the basic structure for solution development, there are other methods for generating ideas that involve harnessing the diversity present in our surroundings. This means tapping into the rich diversity that surrounds us daily, whether it is through the diverse skills and experiences of our team members or drawing inspiration from nature's diversity.

Tapping into the diversity of our teams involves fostering collaboration across various departments or disciplines and allows us to approach challenges from multiple angles that introduces diverse perspectives that can bring unique approaches to how individuals perceive and solve problems, consequently inspiring numerous innovative solutions. Cultures around the world have distinct values, traditions, and ways of thinking shaped by their history, environment, and societal norms. These differences influence how people perceive challenges and opportunities. A problem that seems straightforward in one culture might be viewed as multifaceted in another, leading to different insights and potential solutions.

For example, in a post from the Youth Networking Business Corporation organization (YNBC, 2021), it is noted how in certain Asian cultures like in Japan, waste management takes on a holistic approach that blends cultural norms and technological advancements, contrasting with other cultures that primarily address it as an environmental sustainability and resource conservation issue. In these Asian cultures, where waste management is viewed more as a social responsibility, solutions tend to be more diverse, focusing on individual behavior and where issues are initially framed (defined) to encourage social accountability actions, such as conducting community clean-up drives, composting, and awareness campaigns. In contrast, in countries where waste management is seen as more of an environmental issue, the solutions tend to be centered around building recycling infrastructure.

To foster divergent thinking and nurture curiosity throughout the ideation process requires us to create an environment that embraces and thoroughly explores every idea through the promotion of open dialogue and valuing contributions from all team members. We achieve this through inclusive leadership, which involves cultivating an environment where team members feel empowered to share ideas without fear of judgment.

As the Harvard Business Review noted (Jain-Link et al., 2020), fostering an inclusive environment at all levels of the organization involves valuing authenticity over conformity. It means recognizing that a diverse range of presentation and communication styles can contribute to workplace success. By creating such an environment, we not only support the well-being of our team members but also stimulate innovation within a diverse workforce.

Many tools have been developed to assist with Design Thinking in stimulating innovation, such as those included in the design toolkit developed by Liedtka and Ogilvie (Liedtka & Ogilvie, 2011): visualization, journey mapping, value chain analysis, mind mapping, brainstorming, concept development, assumption testing, rapid prototyping, customer co-creation, and learning launch. These tools, as well as other activities and exercises, can be utilized as needed within our ergonomics processes to leverage our team's diversity, fostering an environment conducive to creativity and innovation. Let's review a few of those and how they can be applied in our initiatives such as solution development events (e.g., kaizens).

Divergent Thinking Activities

We've all seen them and may have taken part in—creative exercises where teams, fueled by sometimes intense friendly competition, scramble to build the tallest, most stable structure equipped with little more than tape and uncooked spaghetti or balloons. While hilarity almost always ensues, the bigger outcome of these types of exercises is to get the participants to work together to think creatively about a problem at hand. These exercises are like mini-laboratories where we encourage teams to tackle problems in inventive ways.

What's interesting is how diverse experiences and perspectives come into play. Someone might bring in an idea from their hobby, and suddenly it sparks a whole new approach to the challenge the team is faced with. That's where divergent thinking comes in. The goal is to unleash a flood of ideas by tapping into the varied backgrounds and viewpoints of team members. It is about exploring multiple solutions and breaking away from the usual ways of thinking.

In the kaizen events that I facilitate, I like to do a quick exercise of getting the participants to look at a simple object, say a spoon or brick, and have them write down as many potential uses for it. It starts off straightforward, just listing practical uses. Then, I throw in a twist: Crazy ideas are fair game, and a prize will be given for the most novel idea. That's when you really see the brain gears start to turn!

After the timer runs out, we celebrate both the quantity of ideas generated (because fluency matters!) and the most out-of-the-box use (because creativity shines here!). Then comes the debrief: What challenges did they face? How did they feel when they were encouraged to think wildly? It is all about getting them comfortable with unconventional thinking and setting the stage for an open-minded approach throughout the event. When you mix diverse perspectives and give people permission to think outside the box, you unlock incredible creativity and problem-solving potential.

Emphasize "Be Curious" during Observations

When it comes to problem-solving, divergent thinking brings in a wealth of creative ideas, but maintaining an open-minded approach during observations is equally crucial. When we're out on the floor or in the field during the observation phase of the kaizen (my favorite part), I stress to the team that we're not just observing what is immediately apparent on the surface. We need to dig deeper, looking for those subtle issues and underlying causes that could pose safety risks or create operational inefficiencies (or negatively impact employee well-being, increase defects, etc.). It is essential for us as observers to constantly question the reasons behind what we see. Understanding the "why" behind activities is key—whether they stem from upstream decisions or downstream consequences. This deeper understanding helps us uncover potential hazards that may not be obvious at first glance.

We must also acknowledge that the way work is prescribed (on paper) isn't always how it is actually carried out (in practice). There are often valid reasons why tasks are performed a certain way, even if they involve some level of risk acceptance. Recognizing this discrepancy between theory and practice allows us to make more informed observations and, ultimately, more effective and sustainable solutions.

Ensure Plenty of Opportunity to Build and Test (and Build and Test Some More)

Another thing that is a staple when I facilitate kaizen events is that I always make sure to have a variety of modeling clay and craft materials readily available. It is not just because I enjoy the nostalgic scent that brings me back to childhood—it is because these tangible tools are incredibly effective in demonstrating ideas and prototypes. They help to visualize concepts in a way that words alone often can't, making it easier to convey and refine new solutions. Plus, let's be honest, there's something universally appealing about getting hands-on with glue and popsicle sticks because we're all still kids at heart!

That's why I strongly recommend having a well-stocked "craft box" as a staple for any kaizen event. Imagine it filled with tape, scissors, construction paper, pipe cleaners, and plenty of modeling clay—tools that can turn abstract ideas into tangible prototypes. It is not just about crafting a specific idea; it is about fostering creativity and honing divergent thinking skills.

Have a "Yes, and…" Mindset

Earlier, I emphasized the significance of building on each other's ideas. It is essential not to dismiss ideas prematurely because adopting a "yes, and…" mindset can turn seemingly unattainable concepts into feasible solutions or even spark entirely new avenues of exploration. By creating an environment where every idea is valued, we cultivate a culture where there are "no bad ideas."

In practical terms, "yes, and…" means accepting and expanding upon ideas rather than immediately shutting them down. Even if an idea seems far-fetched at first, exploring it further could reveal hidden potential or inspire innovative approaches. This approach encourages team members to contribute freely without fear of judgment, fostering creativity and collaboration. Inclusion in idea generation often leads to more innovative solutions in problem-solving.

Expanding on this, the "yes, and…" mindset is not just about agreement but about building a collaborative atmosphere where every contribution is seen as a stepping stone to something greater. It encourages active listening and respect for each other's input, which can significantly enhance team dynamics. When team members feel their ideas are valued, they are more likely to engage deeply and think creatively, leading to breakthroughs that might not have been possible in a more restrictive environment. This approach not only boosts morale but also drives innovation, making it a powerful tool for any team aiming to achieve exceptional results.

Other Brainstorming Tips

Brainstorming can be approached in many ways—from spontaneous idea generation with sticky notes (or brainwriting) to using advanced tools like artificial intelligence. However, it is just as easy to overcomplicate the process to the point where it becomes ineffective. Personally, I believe in keeping brainstorming simple, with clear guidelines to provide some structure, offering "freedom within the framework." Here are my key guidelines:

- **Respectful Communication**: Set a rule that when someone is speaking or sharing an idea, they have everyone's full attention. In the spirit of camp "spirit sticks" (or at least what I've seen in movies 😊), I use a stress ball as a speaking totem—whoever holds it gets to speak. Whatever you choose, make sure it is soft since it will likely get tossed around.
- **Allow Some "Alone" Time**: When holding a brainstorming session, be sure to allow time (2–5 mins) for the participants to think of ideas on their own (preferably in silence) prior to coming back to the group for discussion and collaboration.
- **Adopt The "Yes, and…" Mindset**: To foster a non-judgmental atmosphere, encourage participants to respond with "yes, and..." even if they're offering constructive criticism. This approach ensures that all contributions are respected and keeps the flow positive while also allowing room for valuable feedback.
- **Everything Goes during Ideation**: In the early stages, all ideas are welcome, no matter how unconventional they may seem. Don't worry; ideas will be filtered and prioritized later. You never know, an "out there" idea could spark a new direction for other solutions.
- **Money Is No Object during Ideation**: At this stage, encourage participants to think freely without worrying about costs. I've often had people hold back ideas because they thought they might be too expensive. This restriction should be lifted to ensure creativity flows without limits.
- **Keep It Fun and Encourage Movement**: It wouldn't be an ergonomics event if we didn't encourage movement (the old ergo adage of "the best ergo posture is the next ergo posture" feels apt here). In fact, low-intensity movement such as walking has been shown to enhance creativity (Oppezzo & Schwartz, 2014)—so make your event space one that's conducive to

movement and encourage it! I have recently started to incorporate fidget toys in my events and let me tell you, you haven't seen excitement until you see grown adults rummaging through a bag full of toys! Just try to stay away from the ones that make noise (trust me on this).

- **Encourage Full Participation**: Diversity plays a crucial role in generating ideas that address challenges faced by workers daily, shaped by their unique experiences and perspectives. However, for the brainstorming session to be truly effective, it requires input from everyone. As a facilitator, it is essential to ensure that all participants contribute and feel valued for their individual perspectives.

As with all aspects of this book, it is essential to tailor these processes to fit your specific team(s). By involving them in shaping the structure of your brainstorming sessions, especially the behavioral guidelines for engagement, you'll create a process that stands a much better chance of being effective.

Addressing Creative Anxiety

Imagine you're sitting in the famed "C-Suite" among the executive leadership team, invited to contribute to the development of a new employee experience strategy. Kudos for getting a seat at the table for this high-stakes meeting! But then, out of nowhere, you're asked to share your thoughts on how to enhance the employee experience right on the spot. No pressure, right? They simply want to make it interactive and kickstart the process with a quick workshop. But as the spotlight turns to you, your mind goes blank. If only you had the luxury of a day to gather your ideas, you'd be able to impress them with a comprehensive 10-page portfolio, complete with detailed implementation plans. But instead, you're frozen, struggling for words, feeling the sweat trickle down as you're doing your best goldfish impression ("ah ah ah ah"). Just as you're about to sink into the discomfort, someone else jumps in, graciously diverting attention away from you.

Yeah, creative anxiety is real (Daker et al., 2019). It is a form of anxiety that is specific to thinking creatively and it can show up at the most inopportune times. And it is not uncommon. Women are more likely than men to suffer from this form of anxiety (Georgetown University, 2019)—to the point that many women undervalue their creative abilities, as well as leadership abilities, despite being scored higher than men for these competencies (Zenger, 2017). This anxiety can lead women to shy away from creative problem-solving roles (i.e., STEM) and/or suffer through their career (Daker et al., 2019). With creative anxiety, similar to the anxiety you see with public speaking or stage fright, the acute stress from the situation (e.g., being put on the spot to create) can lead to the sympathetic nervous system going haywire—when the physical reactions such as freezing up, sweating, heart palpitations, etc. become such a distraction that you're disengaged from the task at hand.

Now, imagine your team experiencing this exact same pressure during your next kaizen event. I know it has happened to me—and I like to think of myself as a very creative person! But for me, I'm more comfortable sitting with an issue for a bit (I like to call it "marinating") to get my thoughts organized. However, there are times when we need to be spontaneously creative. It is a good life skill to have. And to be fair, thinking creatively on the spot is what I'm asking of the individuals that are participating in my kaizen events. But the good news (yay!) is that acute stress is temporary, and we can train ourselves to get more comfortable with the stress—essentially train your nervous system to better manage these physical reactions… or, as others say, "train your creativity muscles."

I've found that incorporating creative exercises into our work can be immensely valuable. For instance, in one group that I worked with, we would periodically hold "Creative Gym" sessions. These sessions were spontaneous opportunities designed to stretch our creative muscles. During these sessions, we might be tasked with writing a children's story using three random words or crafting a cardboard masterpiece inspired by a specific theme. Initially, it can be a bit daunting, especially if you don't see yourself as particularly creative. However, the beauty of creativity lies in its ability to be nurtured and developed, much like any other skill—it simply requires practice!

These activities do more than just add an element of fun and energy to the kaizen process. They also encourage participants to think outside the box and explore unconventional solutions. This approach fosters a culture where innovation flourishes, and everyone feels empowered to contribute their unique perspectives. By engaging in these creative exercises, we not only enhance our problem-solving capabilities but also build a more inclusive and dynamic environment where every idea is valued and explored.

These creative sessions also serve as a powerful reminder that creativity is not a fixed trait but a skill that can be honed over time. They help break down the barriers of self-doubt and open up new avenues for innovative thinking. As a result, participants become more confident in their creative abilities, leading to a more vibrant and collaborative atmosphere. This, in turn, promotes a culture of continuous improvement and innovation, where every team member feels inspired to bring their best ideas to the table.

Another development activity that could be considered is participating in improv. Yes, improv—as in improvisational theater. You might be thinking that this suggestion is a bit out there, but hear me out. Improv can be an excellent way to boost creativity, enhance collaboration skills, improve communication, build confidence, and develop the ability to think on your feet—all essential traits for fostering a creative mindset. And it has been shown to be helpful for group brainstorming (Gerber, 2009).

At first, the idea of standing in front of a group of people, whether they are familiar faces or complete strangers, and performing something as random as a T-Rex going out on a night on the town impression might seem daunting. It might even cause a bit of anxiety. However, the skills you gain from such activities are invaluable.

Improv exercises can push you out of your comfort zone and encourage you to embrace spontaneity. This can lead to a significant increase in creative thinking as you learn to generate ideas quickly, without overthinking (which, honestly, is what I think trips us up the most when it comes to solution development). Additionally, improv fosters a strong sense of teamwork and collaboration because participants must listen to each other, building upon the diversity of ideas with "yes, and…", and work together to create a cohesive scene. These skills are directly transferable to any team-based work environment, where effective communication and collaboration are key to success. Confidence is another significant benefit of participating in improv. By regularly stepping out of your comfort zone and performing in front of others, you can build resilience and self-assurance and cultivate the ability to think on the spot. Improv exercises can help train your brain to be more flexible and responsive, enhancing your problem-solving capabilities by teaching you to quickly adapt to new information and come up with innovative solutions.

So, consider incorporating these creative muscle-building activities into your and your team's development and watch the ideas flow.

Create An Environment That Captures (and Follows Up On) Ideas

Creating an environment that not only captures ideas but also effectively follows up on them is another critical element for fostering innovation and continuous improvement within any organization. In fact, the follow-up aspect is perhaps the most crucial part of the process, as nothing disengages employees faster than the feeling that their ideas are being ignored.

Creating a culture that values idea generation and follow-up involves several key elements. For starters, it is essential to establish clear channels for idea submission and ensure that these channels are accessible to all employees. It is also important to have a structured process for reviewing and evaluating submitted ideas. This process should be transparent and involve input from various stakeholders to ensure that all ideas are given fair consideration. And as mentioned earlier, the follow-up is critical. Once an idea is submitted—whether selected for implementation or not—it is very important to communicate the progress and outcomes back to the individuals (the specifics as related to their idea) and to the organization as a whole (individual details if agreed upon by the idea owner; otherwise, general aggregate information). This not only recognizes the contributions of the individuals who submitted the ideas but also encourages others to participate in the future. Doing so also shows support from leadership, which is needed to foster a culture of continuous improvement.

Fortunately, there are several effective ways to integrate the collection and follow-up of ideas into any ergonomics process:

- **Suggestion Boxes**: Placing suggestion boxes in strategic locations allows employees at all levels to submit their ideas, anonymously if they prefer. This simple yet effective method encourages participation and ensures that everyone has an avenue to voice their contributions to the organization's growth.
- **Software Solutions**: Utilizing specialized idea collection platforms (several are commercially available) provides a digital platform for idea submission, collaboration, and tracking. These tools create a structured and streamlined

platform for capturing ideas, facilitating communication and transparency, making it easier to manage and follow up on ideas in real-time.

- **Observation, Audits, and GEMBA Walks**: Actively engaging in observation, audits, and GEMBA walks ("GEMBA" is a Japanese term meaning "the real place"; it is a lean manufacturing process of observing where the work is being done) encourages frontline employees to identify opportunities for improvement firsthand. This approach not only captures immediate insights but also demonstrates management's commitment to listening to and acting on feedback from the ground level.
- **Integration into Daily Operations**: Implementing mechanisms where employees can post ideas and suggestions in a centralized location visible to all, integrates idea generation into daily operations. This constant visibility keeps innovation and accountability at the forefront of organizational culture.
- **Internal Competitions**: Initiating internal competitions, similar to what can be seen with the Applied Ergonomics Conference's (AEC) Ergo Cup competition, can motivate employees to innovate by offering recognition and rewards for outstanding ideas. These competitions inspire friendly competition and highlight the value of creative problem-solving within the workforce.
- **Employee "Labs"**: Establishing an "Employee Lab" in collaboration with company departments involved in cultivating employee relations creates a dedicated space for brainstorming, testing prototypes, and refining ideas. This initiative fosters a collaborative environment where cross-functional teams can work together to develop and implement innovative solutions.
- **Annual Employee Surveys**: Conducting regular employee surveys focusing on innovation and idea generation provides valuable feedback on the effectiveness of current processes and identifies areas for improvement. It also serves as a platform for employees to express their opinions and contribute fresh ideas.

DERIVING INSPIRATION FOR SOLUTIONS FROM THE DIVERSITY IN NATURE

I like to think of myself as a bit of an inventor. Back in the day—brace yourself for a nerdy confession—I would carry around a notebook filled with ideas for "future inventions" because I was constantly dreaming up ways to improve things or create new solutions. It is a trait I inherited from my mom, who was always tinkering around the house to fix inefficiencies. She even claims to be the original inventor of the "lazy Susan" (ok, mama!), and we still use the one she made from scrap materials.

Driven by a desire to solve everyday problems, I often turned to nature for inspiration. I might have been fascinated by biomimicry long before I even knew it had a name! I would watch ants build their mounds and marvel at how they managed to carry what seemed like enormous loads for their size. Observing my cats, I was amazed by their ability to twist and stretch their bodies to move with such grace.

And I was equally impressed by the gecko's ability to climb walls effortlessly thanks to its sticky toes. These natural wonders were my early sources of inspiration, sparking ideas for practical ideas rooted in the principles of nature.

And why not look to nature? All you have to do is look at how nature, when faced with new challenges from ecological changes, finds ways to innovate for pure survival. Such as coral reefs develop the ability to form symbiotic relationships with more heat-tolerant algae to survive in warmer waters due to increased ocean temperatures (Zandonella, 2016) or naked mole rats evolving to adapt their metabolism to fructose when oxygen levels are extremely low due to their underground environments becoming harsher with low oxygen levels and high levels of carbon dioxide (Amoroso et al., 2023). Beyond survival, due to edge effects, some species are found to exhibit creativity and innovativeness in transitional areas where one or more ecosystems combine, called ecotones—such as when land and water meet, or even the encroachment of urban developments into woodlands (Potts et al., 2016). For example, bears are becoming innovative and very resourceful by learning how to open car doors (Solis, 2024)!.

The COVID-19 pandemic came with many lessons we were forced to learn. An example of this is how much we could learn (or could have learned) from nature in how to utilize its inherent diversity to adapt and be resilient. In Jamie Miller's video "ADAPTIVE CYCLE | COVID-19 from an ecological perspective" (Miller, 2020), he explains how ecosystems in nature aim for high diversity and complexity because it enhances their resilience to changes, such as pest invasions, which would have less impact on a diverse system compared to a monoculture. This diversity fosters innovation through cooperative competition and resource sharing. Historically, we've seen how innovations, such as the transition from the Model T to modern cars and the rise of electric vehicles, demonstrate how systems evolve to meet new challenges. Clinging to outdated systems or resisting change can lead to inefficiencies and potential collapse. As change is inevitable, organizations, and individuals must decide whether to be disrupted by it or to become disruptors themselves.

We can develop those innovative solutions through the study of biomimicry. Biomimicry is an innovative approach that seeks sustainable solutions to human challenges by emulating nature's patterns and strategies, which are derived from nature's diversity. And a fascinating approach it is! I find it intriguing to see how nature employs its own version of kaizen to identify and leverage its strengths in the face of adversity (again, the bears have learned to open car doors!).

According to Biolearn (Biolearn, n.d.), a biomimicry website for schools, nature banks on its diversity. It recognizes that a more diverse ecosystem is more resilient and less vulnerable. The greater the diversity, the greater the opportunity for adaptation. In Richard Dawson's blog post for WildHub titled "Biomimicry and Nature-Inspired Solutions," (Dawson, 2020) he described how nature is a model for sustainability due to its diversity advantage, calling it an "insurance policy." As nature has learned to recycle waste more efficiently, using renewable power from the sun, it is more resilient to sudden changes, more adaptable over time to new conditions, and can self-regulate through constant feedback.

But why is this important to us as ergonomics practitioners? Nature is often considered the best teacher, and for ergonomics practitioners, it can be a valuable source of inspiration for solving workplace and product design challenges. For starters, biomimicry can offer unique insights for ergonomics practitioners by using the understanding of how nature has evolved successfully over millions of years and using that knowledge as inspiration by emulating nature's designs and processes. By studying how ecosystems handle extreme conditions and challenges, designers and engineers can develop innovative solutions inspired by nature—solutions that can enhance comfort, safety, and efficiency—with the added benefit of promoting the use of sustainable materials and processes using nature's efficient systems.

Biomimicry is becoming increasingly prevalent across various fields, including ergonomics, to develop solutions aimed at improving the functionality and safety of products and processes. Here are some notable examples of biomimicry in ergonomics:

- **Robotic Arm**: Inspired by the smooth and flexible movements of an elephant's trunk, a robotic arm has been designed to move and grip objects with precision and ease (Mitchell, 2023).
- **Moisture-Wicking Textiles**: Lightweight textile materials that mimic the reaction of pinecones to humidity, providing effective moisture management (Chatterjee et al., 2022).
- **Head Protection**: Industrial head protection gear modeled after the structure of the pomelo fruit, known for its impact resistance (BMW, 2017).
- **Knee Pads**: Improved knee pads inspired by the protective shell structure of an armadillo, offering enhanced durability and flexibility (Islam et al., 2021).
- **Hearing Aids**: Advanced hearing aids designed based on the corrugated design of cricket wings, which improve sound clarity and directionality (Malkin, 2016).
- **Ultrasonic Cane**: A cane for the visually impaired that uses echolocation capabilities similar to those of bats, providing better navigation and obstacle detection (Eck, 2013).
- **Enhanced Air Conditioning:** Bionic-inspired counter-current heat exchanger approach inspired by the blood flow in penguin feet (Gregory et al., 2018).
- **Velcro**: Perhaps the most famous example, hook and loop material (Velcro) was inspired by the way cockleburs stick to fur and clothing (Velcro, 2020).

For ergonomics practitioners, incorporating biomimicry into our problem-solving efforts can offer new perspectives and lead to groundbreaking innovations. Exploring how nature addresses issues like energy conservation and load distribution could provide valuable insights for tackling common ergonomics-related challenges. We should keep the idea of nature to develop innovative solutions on our solution development radar and in our toolkit. There are several resources for biomimicry, but to get you started, here are a few I would recommend:

- Ask Nature (Asknature.org)
- Biomimicry Institute (biomimicry.org)
- The Center for Biologically Inspired Design (CBID) at Georgia Tech (cbid.gatech.edu)
- Science Friday: Biomimicry in Action blog (sciencefriday.com)
- Learn Biomimicry's blog (learnbiomimicry.com)
- Interesting Engineering blog on engineering inventions inspired by nature (https://interestingengineering.com/science/biomimicry-9-ways-engineers-have-been-inspired-by-nature)

DIVERSIFYING WHERE WE FIND SOLUTIONS

As highlighted by the exploration of diversity in nature, inspiration for ergonomics solutions can often come from unexpected sources. While it is important to leverage our existing knowledge and experience, seeking out unconventional ideas can lead to groundbreaking solutions. This might involve looking beyond traditional ergonomics domains to discover innovative approaches that address challenges in novel ways.

Because ergonomics touches virtually every aspect of our lives, so does the potential for inspiration. Stay open to discovering new sources of insight, and you may find transformative ideas that enhance both functionality and comfort in unexpected ways. Here are some recommendations—or sources of inspiration, if you will—for places, both common and uncommon, to look for solution ideas:

- **Idea Banks**: Explore a variety of idea banks that focus on ergonomics and related topics, which can be treasure troves of information on existing solutions and best practices. Resources such as the Washington State Industrial Safety and Health Act, OSHA e-tools, and the Job Accommodation Network (JAN) offer valuable insights and suggestions for addressing specific ergonomics-related risk factors and challenges.
- **Disability Communities/Assistive Technologies**: Investigate accessible and assistive technologies designed for individuals with disabilities, as these solutions can often be adapted for work-related contexts. These communities are rich in creativity, frequently needing to innovate in order to navigate a world not necessarily designed for them. Their ingenuity offers valuable insights and can help uncover barriers that might go unnoticed due to our biases. For instance, adaptive keyboards designed for those with limited dexterity could enhance comfort and efficiency for a broader range of users. Similarly, closed captioning, now commonly used in virtual meetings, has proven to improve meeting effectiveness for everyone.
- **Hobby Blogs and Trade Associations**: Explore hobby blogs and trade associations related to specific industries or interests. These sources can offer creative solutions to common issues faced within particular hobbies or fields. For instance, a blog dedicated to woodworking might share innovative tool designs that could be applied to other areas requiring similar manual tasks.

- **Conferences**: Attend conferences and review entries from events like the AEC Ergo Cup, the National Ergo Expo, or HFES Conference. These conferences showcase cutting-edge solutions, ideas, and research from various industries, providing attendees with a rich source of inspiration for developing solutions. Engaging with professionals from around the world at these events can also broaden your perspective and introduce you to innovative practices and technologies.

Assistive Technology & Innovation

My colleague, Gary Downey (featured in Chapter 8), described his work with assistive technology as the "X-Games of ergonomics," or extreme ergonomics. Having witnessed firsthand the creativity involved in developing solutions for individuals with disabilities, I highly encourage all ergonomics practitioners to connect with local organizations focused on assistive technology.

For example, centers like The Charles Lea Center (https://charleslea.org/) and University of South Carolina Assistive Technology Center (https://www.sc.edu/study/colleges_schools/medicine/centers_and_institutes_new/ccente_for_disability_resources/assistive_technology/) offer services and solutions that promote accessibility, independence, and improved quality of life for individuals with disabilities. These centers cover a broad spectrum of innovations, from mobility aids to communication devices and daily living tools. The ingenuity behind these developments is inspiring and presents valuable opportunities for learning *and* potential partnerships.

When faced with the challenge of discovering innovative solutions to workplace problems identified by you and your team, it is essential to apply the strategies and tips we've discussed to ignite creativity. Leverage these approaches to uncover fresh, inventive ideas that can address the issues effectively. And utilize these strategies to motivate and stimulate creative thinking, and drive meaningful improvements and innovations. Encourage both yourself and your team to explore various sources of inspiration, including unconventional ones, and adopt a mindset open to new perspectives. By doing so, you can foster an environment where creativity thrives, leading to breakthrough solutions that not only solve current problems but also enhance overall workplace efficiency and the worker experience.

REFERENCES

3M. (n.d.). About us. Post-it®. Retrieved November 24, 2024, from https://www.post-it.com/3M/en_US/post-it/contact-us/about-us/

Amoroso, V. G., Zhao, A., Vargas, I., & Park, T. J. (2023). Naked mole-rats demonstrate profound tolerance to low oxygen, high carbon dioxide, and chemical pain. *Animals (Basel)*, *13*(5), 819. https://doi.org/10.3390/ani13050819

Attle, B. (2024, February 19). Are we experiencing a sea change in intrapreneurship? *World Economic Forum.* https://www.weforum.org/stories/2024/02/the-changing-face-of-intrapreneurship/

BioLearn. (n.d.). BioLearn: Learning for life. https://biolearn.eu/en

BMW Group. (2017, June 29). Inspired by nature: New body protection for BMW employees. *BMW Group Communications*. https://www.bmwgroup.com/content/dam/grpw/websites/bmwgroup_com/responsibility/downloads/en/2018/2017-BMW-Group-Inspired-by-nature-englisch.pdf

Chatterjee, K., Wei, S., & Ghosh, T. (2022). Chapter 11 - Smart textiles through bio-inspired innovation. In M. Eggermont, V. Shyam, & A. F. Hepp (Eds.), *Biomimicry for materials, design, and habitats* (pp. 291–335). Elsevier. https://doi.org/10.1016/B978-0-12-821053-6.00020-5

Daikoku, T., Fang, Q., Hamada, T., Handa, Y., & Nagai, Y. (2021). Importance of environmental settings for the temporal dynamics of creativity. *Thinking Skills and Creativity, 41*, 100911. https://doi.org/10.1016/j.tsc.2021.100911

Daker, R., Cortes, R., Lyons, I., & Green, A. (2019). Creativity anxiety: Evidence for anxiety that is specific to creative thinking, from STEM to the arts. *Journal of Experimental Psychology: General, 149*(1), 10–23. https://doi.org/10.1037/xge0000630

Dawson, R. (2020, November 11). Biomimicry and nature-inspired solutions. *Wildhub*. https://wildhub.community/posts/biomimicry-and-nature-inspired-solutions

Eck, A. (2013, October 23). Bat-inspired tech could help blind people see with sound. *NOVA*. https://www.pbs.org/wgbh/nova/article/bioinspired-assistive-devices/

Georgetown University. (2019, June 11). Psychology research team identifies creative anxiety. *College of Arts & Sciences, Georgetown University*. https://college.georgetown.edu/news-story/psychology-research-time-identifies-creative-anxiety/

Gerber, E. (2009). Using improvisation to enhance the effectiveness of brainstorming. In *Proceedings of the SIGCHI Conference on Human Factors in Computing Systems (CHI '09)* (pp. 97–104). Association for Computing Machinery. https://doi.org/10.1145/1518701.1518718

Gilbert, S. J., Zamenopoulos, T., Alexiou, K., & Johnson, J. H. (2010). Involvement of right dorsolateral prefrontal cortex in ill-structured design cognition: An fMRI study. *Brain Research, 1312*, 79–88. https://doi.org/10.1016/j.brainres.2009.11.045

Gregory, P., Karthik, D., C, G., Vignesh, & Kumar, G. (2018). Biomimicry for innovative air-conditioning. *International Journal of Advance Research and Innovation, 6*(2), 102–107. ISSN 2347-3258.

History.com Editors. (2020, July 1). U.S. Post Office introduces ZIP codes. *History*. https://www.history.com/this-day-in-history/u-s-post-office-introduction-of-zip-codes

Islam, M. K., Hazell, P. J., Escobedo, J. P., & Wang, H. (2021). Biomimetic armour design strategies for additive manufacturing: A review. *Materials & Design, 205*, 109730. https://doi.org/10.1016/j.matdes.2021.109730

Jain-Link, P., Kennedy, J. T., & Bourgeois, T. (2020, January 13). 5 strategies for creating an inclusive workplace. *Harvard Business Review*. https://hbr.org/2020/01/5-strategies-for-creating-an-inclusive-workplace

Lechner, C., & Gudmundsson, S. V. (2014). Entrepreneurial orientation, firm strategy and small firm performance. *International Small Business Journal, 32*(1), 36–60. https://doi.org/10.1177/0266242612455034

Liedtka, J., & Ogilvie, T. (2011). *Designing for growth: A design thinking toolkit for managers*. Columbia University Press.

Malkin, R. (2016, September 1). Insects are helping us develop the future of hearing aids. *The Conversation*. https://theconversation.com/insects-are-helping-us-develop-the-future-of-hearing-aids-64619

Meyer, K. (2019, July 19). Margaret Hamilton led NASA's software team that landed astronauts on the moon. *Smithsonian Magazine*. https://www.smithsonianmag.com/smithsonian-institution/margaret-hamilton-led-nasa-software-team-landed-astronauts-moon-180971575/

Miller, J. (2020, April 9). ADAPTIVE CYCLE | COVID - 19 from an ecological perspective [Video]. YouTube. https://www.youtube.com/watch?v=2Oy9LyVThlg

Mitchell, M. (2023, October 26). Introducing the trimmed helicoid, a soft robot arm that's based on an elephant's trunk. *TechXplore*. https://techxplore.com/news/2023-10-trimmed-helicoid-soft-robot-arm.html

Oppezzo, M., & Schwartz, D. L. (2014). Give your ideas some legs: The positive effect of walking on creative thinking. *Journal of Experimental Psychology: Learning, Memory, and Cognition, 40*(4), 1142–1152. https://doi.org/10.1037/a0036577

Porumboiu, D. (2021, October 14). What is intrapreneurship and how can it support corporate innovation? *Hype Boards*. https://www.viima.com/blog/intrapreneurship

Potts, J. R., Hillen, T., & Lewis, M. A. (2016). The 'edge effect' phenomenon: Deriving population abundance patterns from individual animal movement decisions. *Theoretical Ecology, 9*(2), 233–247. https://doi.org/10.1007/s12080-015-0283-7

Solis, N. (2024, September 21). Bears have learned how to open doors in the Sierra Nevada. *LA Times*. https://www.yahoo.com/news/bears-learned-open-doors-sierra-100005260.html

Stanford University Institute of Design. (2016). Welcome to the virtual crash course in design thinking. http://dschool.stanford.edu/dgift/

Urban, B., & Wood, E. (2015). The importance of opportunity recognition behaviour and motivators of employees when engaged in corporate entrepreneurship. *Journal of Business Economics and Management, 16*(5), 980–994. https://doi.org/10.3846/16111699.2013.799087

Velcro Companies. (2020, July 27). A mind-blowing biomimicry example. *Velcro Companies*. https://www.velcro.com/news-and-blog/2020/07/a-mind-blowing-biomimicry-examples/

Youth Networking Business Corporation (YNBC). (2021, June 15). Revolutionizing waste management: Asia's innovative solutions for a circular economy. *Youth Networking Business Corporation*. https://www.ynbc.org/post/revolutionizing-waste-management-asia-s-innovative-solutions-for-a-circular-economy

Vries, H. B., & Lubart, T. I. (2019). Scientific creativity: Divergent and convergent thinking and the impact of culture. *Journal of Creative Behavior, 53*(2), 145–155. https://doi.org/10.1002/jocb.184

Waidelich, L., Richter, A., Kölmel, B., & Bulander, R. (2018, June). Design thinking process model review. In *2018 IEEE International Conference on Engineering, Technology and Innovation (ICE/ITMC)* (pp. 1–9). IEEE. https://doi.org/10.1109/ICE/ITMC.2018.8506838

Zandonella, C. (2016, November 2). When corals met algae: Symbiotic relationship crucial for reef survival dates to the Triassic. *Princeton University*. https://www.princeton.edu/news/2016/11/02/when-corals-met-algae-symbiotic-relationship-crucial-reef-survival-dates-triassic

Zenger, J. (2017, April 3). Are women better leaders than men? *Medium*. https://medium.com/the-mission/are-women-better-leaders-than-men-9fb328065b8b

5 DEI and Regulatory Compliance

At its core, ergonomics is about designing with the user in mind, aiming to create solutions that fit the diverse range of physical, cognitive, and sensory abilities found in any population. Global regulations are increasingly embracing the "design for all" mentality that is fundamental to ergonomics. In this way, ergonomics contributes to Diversity, Equity, and Inclusion (DEI) efforts by being integrated into various global regulations tied to DEI strategies.

For example, ergonomics principles are applied to specific workplace standards to ensure that workstations are adjustable and comfortable for people of different sizes and abilities, enhancing productivity and reducing injury risk. These regulations also guide the development of communication materials, ensuring that they are accessible to individuals with varying literacy levels, language backgrounds, and cognitive abilities. We can see how ergonomics is being used to align safety, health, and workplace design with DEI principles.

These global regulations include standards developed by the International Organization for Standardization (ISO), which is an independent international organization. The ISO standards are developed through a collaboration of national standard bodies representing the interests of their respective country members (e.g., Instituto Português da Qualidade for Portugal, National Association of Manufacturers in Mechanics (UNI) for Italy, American National Standards Institute (ANSI) for the U.S., and British Standards Institution for the U.K.). This is done by way of technical committees or subcommittees formed to focus on specific standard topics, be it a process, product, or material. Other relevant regulations specific to ergonomics and DEI outlined here in this chapter also include examples of specific country regulations or government/federal orders. The list is by all means not exhaustive, and ergonomics practitioners should be mindful of regional/local regulations that may impact them and their respective organizations.

EXAMPLE OF WORKPLACE STANDARDS/REGULATIONS

- **ISO 9241**: The ISO 9241 is an international standard that specifically addresses ergonomics and human-computer interaction. Its primary focus is on the design of user interfaces and work environments related to information technology, ensuring they are user-friendly and effective. It covers a wide range of topics related to ergonomics in digital environments, including visual display terminals, software design, and usability. The standard provides guidelines on designing systems that accommodate users' physical and cognitive needs, and it emphasizes the importance of designing systems

DOI: 10.1201/9781032658988-7

that are easy to use and understand. This includes factors like screen readability, control responsiveness, and overall user interaction.

- For example, the ISO 9241 regulation would promote the design of a software application with intuitive icons and menus to enhance the user experience. Or it could ensure that a website's font size and color contrast are accessible to individuals with visual impairments.

- **ISO 45001**: Whereas ISO 9241 focuses on ergonomics and human-computer interaction, ISO 45001 focuses on occupational health and safety management systems. It provides a comprehensive framework for managing health and safety risks in all work environments to create a safer and healthier workplace. Yet, it offers another great example of how ergonomics intersects with DEI within global regulations. For starters, it emphasizes the need to identify and address workplace hazards and risks, including both physical and psychosocial factors. This involves assessing these hazards in the context of a diverse workforce, where individual differences—such as physical capabilities and cultural backgrounds—can influence risk exposure and perceptions of safety. The standard also advocates for a participatory approach to health and safety management, denoting the importance of involving all employees in the design and control of a safe work environment. The premise behind this, as discussed in the chapters covering participatory approaches, is that by engaging a diverse workforce more effectively, the varied needs and concerns of all employees are addressed, thereby enhancing the overall safety and inclusion of the workplace.
  - Clause 7.4 of ISO 45001 further highlights the importance of considering diversity. It stresses the need for effective safety communication by requiring organizations to take into account different cultures, languages, and literacy levels when distributing safety information. This is where ergonomics comes into play: Ergonomics is about designing systems that fit the user, and in this context, it means ensuring that safety communications are accessible to everyone, regardless of their background or language skills. Mitigation examples could involve providing information in multiple languages, using simple and clear language, and incorporating visual aids to support understanding.
- **British Standard 76005 (United Kingdom)**: Another significant example of ergonomics incorporated into regulations directly enhancing DEI is British Standard 76005, which specifically focuses on valuing people through diversity and inclusion. This standard provides a code of practice for organizations, emphasizing the need to integrate DEI principles into their operations and cultures. Ergonomics is a key component here because it involves creating environments and processes that accommodate diverse needs. For example, this standard might guide companies to implement ergonomics processes that support employees with various physical abilities, such as adjustable workstations and accessible design features.

- **Executive Order 14035 (United States)**: In the U.S., Executive Order 14035 (issued in 2021 by the Biden Administration) is another vital piece of regulation that highlights the importance of DEI, specifically within the federal workforce. This order calls for a focus on enhancing diversity, equity, inclusion, and accessibility across federal agencies. The goal is to remove barriers to equal opportunity and provide resources that support a diverse and equitable workforce. Ergonomics contributes to this effort by ensuring accessible and inclusive work design of processes and environments. This could include implementing adjustable workstations, designing accessible office layouts, or providing assistive technology to meet accommodation needs.
- **ISO 30415**: This standard, aimed at Human Resources (HR), provides guidelines for common HR processes designed to promote and assess progress in valuing diversity. It covers areas such as workforce planning, remuneration, recruitment, induction, learning, and development. The standard also aligns with the U.N. Sustainable Development Goals, including those focused on gender equality, innovation, and reducing inequalities. Ergonomics practitioners should stay connected with this standard, as these areas present opportunities for ergonomics processes to make a positive impact (and thus, further demonstrate their value).

OTHER NON-WORKPLACE SAFETY GOVERNMENTAL REGULATIONS

Many countries are adopting standards that require public spaces, such as public parks, buildings, and transportation systems, to be accessible to everyone, including individuals with disabilities. Ergonomics principles within these regulations help ensure that facilities are designed with features like ramps, adjustable seating, and clear signage. For instance, the U.S.'s Americans with Disabilities Act has set a benchmark for accessibility that influences global standards. These regulations not only improve accessibility but also promote a sense of inclusion and respect for all individuals.

The same can be said for ergonomics' role in product design regulations. Take, for example, regulations like the U.S. Consumer Product Safety Improvement Act, the Federal Drug Administration guidelines, the European Union's General Product Safety Directive, or Australia's Australian Consumer Law. These regulations govern the creation, development, distribution, and marketing of consumer products, such as medical devices or cosmetics. Ergonomics principles embedded in these product safety regulations ensure that the products they cover are designed to be usable by a diverse population, including individuals with disabilities. Examples of ergonomics design/usability principles in these regulations include the incorporation of larger buttons, tactile feedback, and adjustable interfaces in smartphones and tablets, making them more accessible to older adults and those with mobility, visual, or auditory impairments. These design considerations further contribute to equitable access to technology and everyday conveniences, aligning with DEI principles.

Beyond product development, ergonomics as a consideration is becoming increasingly relevant in the development of international environmental and sustainability regulations. For example, in the design of sustainable infrastructure projects, ergonomics can help create environments that not only meet ecological standards but also provide safe, comfortable, and inclusive spaces for diverse populations. This might include designing public spaces that are both environmentally friendly and accessible to individuals with different physical needs, thereby promoting a more inclusive approach to sustainability, as laid out in the United Nations' New Urban Agenda (UN, n.d.). This agenda provides the global framework for urbanization to include accessibility and disability inclusion in economic and social urban development, enabling disabled individuals to live independently and participate fully in all aspects of life. Additionally, with regards to Environmental, Social, and Governance efforts, ergonomics is firmly incorporated in the design and development of workspaces that minimize the risk of injuries, but also in the design and development of sustainable processes, products, and buildings. For example, this could include designing office spaces that optimize energy efficiency as well as worker comfort, safety, and operational efficiency.

As demonstrated, the integration of ergonomics into various global regulations and standards not only supports DEI principles but also improves safety, accessibility, and inclusion across multiple sectors. This highlights the value of ergonomics, and by extension us as ergonomics practitioners have on creating a more sustainable and equitable world.

ADDRESSING THE CHALLENGES

Despite the potential benefits standards and regulations provide to an organization in the form of structured methods of improving workplace safety and health, employee well-being, process efficiency, and DEI goals, there are challenges to effective implementation that can mute those benefits. One major hurdle is the complexity of navigating and complying with a wide array of regulations, which can be overwhelming. For ergonomics practitioners with global responsibilities, this challenge is amplified. Not only must we adhere to international regulations, but we must also consider local or country-specific rules. Depending on your location and responsibilities, that could mean you/your process having to manage compliance with potentially overlapping or divergent national, state/province, and local regulations. This necessitates a thorough understanding and interpretation of all applicable regulations, which may vary from one facility to another. And given the potential complexity and ongoing evolution of regulations and standards, this could mean a great deal of resources needed to manage this element of an ergonomics process.

Resistance to change is another potential barrier. This resistance can manifest in multiple ways including operational pushback against implementing potentially costly improvements or employees' reluctance to adopt new systems or processes, particularly any changes related to DEI given the political and social climate associated with DEI efforts.

And beyond the regulatory complexity and potential resistance to changes in the implementation and management of these standards and regulations, developing solutions that effectively accommodate the diverse needs of a diverse working population and measuring the effectiveness of those solutions and designs specifically in meeting DEI objectives present their own set of challenges.

To address these regulatory compliance and implementation challenges effectively (as well as challenges of demonstrating ergonomics' impact), a multi-faceted approach is essential, including education, leadership commitment, effective utilization of technology, cross-functional collaboration, and discussions around future regulations.

Education and Training

First and foremost, education and training play a key role in bridging the gap between understanding the regulations that govern your specific location(s) and implementing them effectively. Ergonomics-related training should not only include discussions on the applicable regulatory requirements for compliance but also delve into the broader connection ergonomics has with the regulations, specifically the DEI principles (and their purpose) embedded in these regulations. It is important to convey the "why" behind these regulations, such as explaining how providing safety communications in multiple languages can enhance overall safety as well as promote inclusion. By emphasizing the fundamental connection between ergonomics and DEI principles, and how they are integrated, individuals can better appreciate the reasons behind these regulations, their impact, and any implemented process changes as a result of them (e.g., the implementation of multi-language safety instructions).

To stay aligned with evolving regulations, it will be important to establish a continuous training plan. Training programs should regularly incorporate updates to keep management and operators (as applicable) informed about new or revised regulations. This may include information on upcoming regulations, along with discussions with employees on how to best adjust operational practices to align with regulatory trends while considering local workforce needs.

Influencing Leadership Commitment

Tailoring the training content designed specifically for each stakeholder (operators, management, engineers/designers) will help to ensure clarity on each stakeholder's role and responsibilities with the standards/regulations. Tailoring the training to address the unique needs and perspectives of each group will help to foster an understanding of how each contributes to the overall process and the importance of their involvement for successful implementation and compliance.

For the leadership team, it is especially important to emphasize the benefits of the standards and regulations, not just from a compliance standpoint but also in terms of the positive impact on organizational success. While legal compliance is certainly a driving factor, it is equally essential to articulate the broader "why" behind these efforts. Their active engagement and visible support are key to securing the necessary resources and ensuring the success of the ergonomics (and safety) processes. Therefore, it will be critical for us to show how the organization's adherence to these standards (through their leadership) can improve safety, productivity, and employee well-being, as well as how the successful implementation of these standards can also meet any organizational DEI goals.

Leveraging Technology

Leveraging technology also plays a significant role in addressing regulatory implementation challenges. Advanced tools, such as risk assessment software or safety management systems, can streamline the implementation of regulations as well as the monitoring of solution effectiveness (particularly as related to quantifying impact to DEI metrics). Risk assessment software can analyze various workplace hazards and identify trends, through predictive analytics, before they arise, enabling proactive measures to be taken. Safety management systems can track incidents, manage audits, and ensure that corrective actions are taken, all while maintaining comprehensive records that document regulatory compliance.

However, going further and utilizing advanced technology can offer innovative solutions to aid in the implementation of ergonomics (and safety) processes, thus enhancing the successful implementation and adherence to standards and regulations. For example, data analytics tools (such as Power BI, Tableau, or SAS) can help ergonomics practitioners (and DEI managers) monitor and evaluate the effectiveness of ergonomics solutions over time through the analysis of metrics such as injury rates, productivity levels, and employee feedback. This can provide valuable insights into how well interventions are performing and where improvements might be needed (e.g., using risk assessment data to identify any gender disparity gaps within jobs). This data-driven approach can help with the continuous improvement of practices as well as help meet any relevant regulatory requirements.

As another example, virtual reality (VR) and augmented reality (AR) could be used to more effectively train employees on ergonomics (and safety) protocols. The VR could be used to create immersive training environments where employees experience realistic simulations of workplace scenarios (with potential risks), allowing them to practice and understand proper techniques in a risk-free setting. The AR, on the other hand, can overlay digital information onto the physical workspace, providing real-time guidance and feedback on ergonomics and safety procedures as employees perform their tasks, making them better prepared to handle real-world challenges.

Lastly, software such as digital human modeling could be used to more effectively design workstations that accommodate a diverse workforce and/or diverse operator needs. These kinds of software packages can enable ergonomics practitioners/engineers/designers to design and modify workstations with more precision, ensuring that the designs better account for a diverse workforce as well as meet any accessibility regulations.

Cross-Functional Collaboration

As ergonomics practitioners, collaboration with different departments within the organization to proactively address and anticipate regulatory compliance concerns is paramount. By working together with these departments, we can stay updated on regulatory requirements and develop cohesive solutions and strategies to meet those requirements. This cross-functional collaboration ensures that our approach to ergonomics and DEI is comprehensive, effectively addresses all relevant compliance issues, and is firmly embedded into the organization's way of doing business.

This could involve engaging with key departments such as:

- **Safety**: Ensure that ergonomics initiatives are aligned with safety regulations, that both safety and ergonomics standards (where applicable) are continuously updated with efforts to meet the standards are prioritized to ensure sufficient resources, and that any solutions or initiatives undertaken to adhere to the standards are done so with a DEI lens.
- **Operations**: Coordinate to incorporate ergonomics and DEI considerations into operational procedures, ensuring that day-to-day activities comply with relevant regulations.
- **Human Resources**: Integrate ergonomics and DEI principles into employee training programs, recruitment processes, and overall workplace policies; additionally, collaborate with key HR stakeholders to align metrics.
- **Procurement/Purchasing**: Ensure that the acquisition of equipment and materials meets ergonomics and DEI standards, supporting compliance within these processes.
- **Legal**: Stay informed about regulatory changes and ensure that all ergonomics and DEI practices and initiatives comply with current laws and regulations.

Providing Input to Shape Future Regulations

One of the most effective ways to truly grasp a regulation is to actively engage in its development process. Participating in industry forums, regulatory review committees, and standard-setting organizations can offer invaluable opportunities for ergonomics practitioners to stay ahead of emerging regulations and shape standards that prioritize both ergonomics and DEI. By joining these groups, we can gain early insights into proposed changes and contribute our expertise to influence the direction of new regulations.

For example, involvement in forums like the International Ergonomics Association, the ANSI, or other relevant industry associations allows ergonomics practitioners to contribute to standards that directly impact workplace ergonomics and diversity. We can provide feedback on draft regulations, share best practices from our own experiences, and suggest adjustments to ensure that the standards support inclusive and accessible work environments. Of course, this will require time and effort, but this proactive involvement can help to ensure that any final regulations are practical, effective, and are aligned with our organization's voice with regards to ergonomics and DEI. This, in turn, could also benefit our ergonomics processes by making us aware of the latest trends and advancements in ergonomics (and DEI), making it a valuable and worthwhile undertaking.

FUTURE TRENDS IN REGULATIONS

Discussions about DEI are increasingly complex, especially in the current political climate (see Part III for more details on the various challenges facing DEI). Despite these challenges, the general consensus is that DEI (as a business strategy) is here to stay. However, the conversation and efforts around it are expected to continue to evolve as organizations and societies around the world become more attuned to the complexities of diverse workforces and communities.

What could this look like?

It could mean more focus on data and DEI-related metrics to show progress. According to insights from Wharton looking into the future of DEI, including perspectives of industry leaders and academics, there is expected to be an emphasis on data-driven initiatives (Basiouny, 2024). This involves collecting and analyzing data to assess the effectiveness of DEI strategies and making informed adjustments based on these insights.

It could also mean that, rather than being treated as a separate or supplemental initiative, DEI will be integrated into other business strategies and objectives to align it more with the core mission and daily operations of organizations. It is also expected that there will be a shift toward incorporating DEI into the routine functions of businesses, ensuring it becomes a fundamental part of organizational culture rather than an afterthought.

Part III delves deeper into these and other future DEI activities within organizations and the role ergonomics can play in these efforts. As the landscape of DEI within organizations evolves, we can expect corresponding changes in regulations, and the organizations we support must be prepared to adapt in order to remain compliant. Therefore, it is essential to recognize the significant impact this evolution will have on shaping our ergonomics processes.

For instance, this may involve further integrating ergonomics into design processes to identify and mitigate physical, cognitive, psychophysical, or psychosocial risks for diverse workforces. This includes promoting psychological safety, managing stress, and addressing work-life balance and mental health issues (discussed further in Part II). As worker well-being and overall health become a growing priority, our profession will take a leading role in these areas, incorporating risk analysis and management to support these objectives.

Additionally, it may mean expanding risk assessments to address emerging risks in a changing and diverse work environment, such as those linked to remote work, flexible arrangements, non-traditional workspaces, and gig economy roles.

Alongside traditional metrics related to safety and operational efficiency, there will likely be a greater emphasis on DEI metrics. As such, we will need to be able to demonstrate progress in these areas, such as monitoring the effectiveness of risk mitigation across different demographic groups.

It could also involve active participation in technical groups and associations that influence regulatory evolution, ensuring that ergonomics practitioners are part of shaping new guidelines.

Finally, as regulations increasingly require diverse perspectives in designing and implementing solutions that affect the workforce, this could lead us to expanding and strengthening our participatory approach, further supporting DEI strategies and initiatives.

In light of these potential changes, ergonomics practitioners must be proactive in preparing for the future. This may include:

- Revising or expanding ergonomics assessments to incorporate psychosocial risks.
- Updating design criteria to cater to a diverse workforce (see Part II for more details and guidance).
- Initiating or continuing the use of a Participatory Ergonomics approach, ensuring diverse voices are included throughout the stages of design, risk identification, analysis, solution development, implementation, and control.
- Expanding ergonomics training and communication to more clearly highlight the DEI elements we impact.
- Enhancing metrics to track DEI improvements, which involves broadening data collection methods, employing new tools or processes, and analyzing trends to develop both reactive and proactive mitigations.
- Ongoing personal training on DEI and regulatory updates to stay current with evolving standards and practices.

As they say, the future is now. This emphasizes the urgency of preparing for upcoming changes, especially in the context of evolving regulatory landscapes and political conversations, both locally and globally, around DEI. As we navigate these changes, it is important to understand that the regulatory landscape is not static. It continuously evolves, influenced by political, social, and economic factors. This dynamic nature requires us to stay informed and adaptable. By doing so, we can foresee the potential impacts on our profession and processes, allowing us to strategically direct our focus and resources to better prepare for these changes.

REFERENCES

Basiouny, A. (2024, June 10). *Is DEI going away? Here's what experts say.* Knowledge At Wharton. University of Pennsylvania. https://knowledge.wharton.upenn.edu/article/is-dei-going-away-heres-what-experts-say/

UN-Habitat. (n.d.). *New Urban agenda.* United Nations Human Settlements Programme. Retrieved May 9, 2024, from https://unhabitat.org/about-us/new-urban-agenda

Part II

Designing for and with DEI

"Where does Ergo live within an organization?"

This question arose during a discussion I had with Rachel Michael, a Certified Professional Ergonomist and a Certified Healthcare Safety Professional with over 20 years of global experience in ergonomics and risk control. We were examining the role of ergonomics in organizations, emphasizing that its scope should extend beyond safety to include operations and design. Rachel emphasized the importance of integrating ergonomics into broader organizational functions, making it a core component of work system design rather than merely a reactive measure for health and safety issues. Although ergonomics has effectively demonstrated its value in enhancing safety, this focus has sometimes limited perceptions of our role. Unfortunately, our profession has been poorly marketed as only addressing health and safety concerns, often reducing ergonomics to just improving physical conditions, such as preventing back pain from poor chair design.

Historically, ergonomics has been closely linked to health and safety, dealing with workstation design and the prevention of musculoskeletal disorders (MSDs). As a result, there is a risk that organizations view ergonomics as a secondary concern, relying on reactive measures based on trailing indicators to address issues. This safety-centric approach often means that interventions occur only after problems or injuries are identified.

However, this focus on safety should not diminish the importance of ergonomics in health and safety programs, initiatives, or departments. The challenge lies in demonstrating that ergonomics encompasses much more than just health and safety. The positioning of ergonomics within organizations can sometimes obscure its broader

DOI: 10.1201/9781032658988-8

impact and potential. Yet, Rachel believes that ergonomics is finally evolving beyond its traditional safety focus to influence broader operational and design aspects.

> If the ergonomics program does not tie into design, it is better described as an 'injury response' program… [It] should be in place first to direct activities such as employee training, hazard assessment, workstation design and equipment selection.

One key area where ergonomics intersects is with DEI, particularly through its role in organizational design. Although the link between ergonomics and general work design—concerned with how work is structured, organized, experienced, and performed (Morgeson and Humphrey, 2008)—is often more apparent, ergonomics (more specifically, macroergonomics) also plays a crucial role in organizational design. Organizational design encompasses the overall structure and framework of the organization, including the arrangement of resources, roles, and processes, and ergonomics contributes significantly by ensuring the work environment supports the organization's goals (Karwowski and Zhang, 2021).

Integrating the principles of ergonomics into organizational design fosters a more comprehensive approach to the employee experience, aligning with DEI goals of removing barriers and supporting employees to contribute fully. Therefore, broadening the scope of ergonomics represents a strategic investment in both employee well-being and organizational success, aligning fully with the objectives and intentions of DEI initiatives.

This captures the core purpose of Part II: *To highlight ergonomics as a proactive role in designing work, processes, products, tools, and the overall worker experience.* As such, it is imperative for us to *design* solutions that cater to a diverse workforce, ensure equity, and foster inclusion to achieve the strategic advantages of enhanced organizational performance and profitability.

DESIGNING FOR A DIVERSE WORKFORCE

Designing for a diverse workforce through ergonomics not only mitigates risks and improves efficiency but also promotes a workplace culture that celebrates differences and maximizes the potential of all individuals. It is about harnessing the power of design to create environments and tools that empower every worker to perform at their best, contributing to both individual and organizational success. While addressing the needs of a diverse workforce is undeniably complex due to the varied challenges of working with people, we are, after all, "problem solvers"—and this is precisely what we do best.

I had the opportunity to watch a very insightful panel discussion about designing for diversity. sparks and honey (sparks and honey, 2023) held a discussion in October 2023 that laid bare the concerns designers have with designing effectively for diversity. The main challenge asked and discussed was how to effectively design for all the complexities that encompass a diverse world and how to do so pragmatically. It is a lofty goal, but can we a*ctually* do it?

In answering the question, the panelists explored the necessity of adopting an inclusive design mindset that considers the varied needs of individuals across different racial, ethnic, gender, and age groups. They emphasized the importance of making designs accessible and welcoming to everyone. A key takeaway was the need for designers to engage with diverse communities throughout the design process. This engagement helps to gain understanding to better address the unique challenges faced by different populations, ensuring that their perspectives are reflected in the final solutions.

Listening to this discussion cemented my perspective on the ergonomics profession. Just as the panel emphasized an inclusive "design for all" approach, ergonomics inherently embodies this philosophy by striving to meet the needs of diverse workers and end users. Our profession is fundamentally about designing work and products that accommodate the complex, varied nature of human beings. Through participatory approaches, ergonomics practitioners engage with workers and end users, giving them a voice in identifying, prioritizing, and solving issues related to their safety, operations, and organizational effectiveness.

Our deep understanding of the human system—encompassing physical, cognitive, and social aspects—positions us uniquely as champions of inclusive design. The skills and insights we bring to the table are needed for ensuring that our solutions not only meet the functional needs of a diverse workforce and/or end-user population but also promote equity and inclusion. Our role as ergonomics practitioners *is* to create environments and products that support the diverse needs of all individuals.

WHAT'S TO COME?

In Part II, my goal is to demonstrate that designing for a diverse workforce through ergonomics does more than reduce risks and improve operational efficiency—it also fosters a workplace culture that values differences and maximizes the potential of all individuals. I aim to highlight how our profession adds significant value to organizations by creating workspaces, tools, processes, policies, products, and, perhaps most importantly, worker experiences that empower everyone to thrive, driving both personal and organizational success.

In Chapter 6, "Ergonomics and the Future of Work," I will explore the inherent diversity in the work environments we design for, including changes in work patterns, locations, and workforce demographics. While ergonomics is often associated with addressing physical differences, such as gender and disability diversity, these aspects will be explored further in later chapters. For example, Chapter 7, "Designing for Gender Diversity" will focus on considering gender-specific physical and societal differences in design, such as the design and selection of personal protective equipment. Chapter 8, "Designing for Diverse Abilities" will emphasize a participatory approach, advocating for ergonomics practitioners to collaborate with individuals with disabilities to better understand their needs and integrate their feedback into design processes, fostering inclusion. Additionally, Chapter 10, "Designing for Varying Generations" will explore design considerations for both older and younger workers.

However, other aspects of diversity also impact how people work, perceive work, and respond to their work environments—all of which ergonomics can influence. Ergonomics, at its core, is about understanding how characteristics such as neurodiversity, gender, race, ethnicity, age, socioeconomic status, and LGBTQ+ identities affect work experiences, and how to apply this understanding in design. This includes both physical design, which accommodates diverse physical needs, and psychosocial design, which addresses how social and psychological factors influence work experiences.

Psychosocial factors are particularly important when designing for diversity, especially when physical differences are not the main concern. These factors involve potential negative impacts on mental and emotional well-being related to issues such as race, ethnicity, culture, socioeconomic status, or LGBTQ+ identity. It is essential to consider psychosocial risks and their role in ergonomics and DEI. While these factors are also discussed in the previously mentioned chapters included in Part II, they will be a central focus in Chapter 9, "Designing for Neurodiversity," where the importance of creating inclusive workplaces for neurodivergent individuals will be explored. Chapter 11, "Inclusive and Equitable Design" will examine the critical role ergonomic design plays in addressing psychosocial factors like stress and job satisfaction, impacting overall worker safety. And in Chapter 12, "Ergonomics and DEI in Communications," I conclude Part II with a review of the various diversity differences and guidance on how to design for them in communications.

Understanding these psychosocial risks is key to creating environments that not only meet physical needs but also support the mental and emotional well-being of a diverse workforce. These considerations will be woven throughout the upcoming chapters, but first, let's take a moment to explore psychosocial factors in more depth before moving into those topics.

PSYCHOSOCIAL RISKS

According to the European Agency for Safety and Health at Work (EU-OSHA), psychosocial risks are issues that can occur at work or carry over from the outside into the workplace and can affect a worker's psychological, social, or physical health (European Agency for Safety and Health at Work, n.d.). EU-OSHA describes these risk factors as:

- **Excessive Demands**: Excessive demands or conflicts in the working environment.
- **Lack of Personal Control**: Having inadequate say over how work is done.
- **Inadequate Support**: Having inadequate support from managers or co-workers.
- **Poor Relationships (Including Harassment)**: Being subjected to unacceptable behaviors (including harassment or violence).
- **Role Conflict or Lack Of Clarity**: Not understanding roles and responsibilities.
- **Poor Management Of Change**: Not being involved and informed in organizational changes.
- **Violence from Third Parties**: Verbal or physical violence, or the threat of violence.

If any element of the work environment leads to a worker having to deal with these risk factors, for example, poor relationships within the worker's department due to poor communication, that risk can contribute to negative health effects such as injuries or illnesses.

However, these risks don't only have the capacity to impact workers' health. When these risks aren't properly managed, they can manifest in high injury rates, poor job performance, frequent absences, high turnover, and a toxic workplace culture. And the impact stretches beyond the company; it can be detrimental to surrounding communities. For example, EU-OSHA estimated a cost of billions of euros to European businesses and societies due to organizational impacts from psychosocial risk factors (Roberts, 2019). And the Canadian Centre for Occupational Health and Safety (CCOHS) cited mental health had an impact on the Canadian economy to the tune of over $50 billion CAD due to healthcare, work disruptions, and lost time (Ferreras, 2016).

Psychosocial Risk and MSDs

As ergonomics practitioners, our job is to understand how workers interact with their work—whether processes, workspaces, or tools/equipment. Psychosocial risks, even though they aren't physical, fall squarely within our scope. We need to assess, control, and manage these non-physical risks just as we do with physical ones. Plus, it is part of our role to educate others about these risks and how to handle them within our ergonomics processes. So, addressing psychosocial risks isn't just a nice-to-have; it is a crucial part of a sustainable ergonomics process that can create and promote a healthy and productive work environment.

As an additional point of note, studies increasingly show that stress from these risk factors does contribute to the occurrence of MSDs (Oakman et al., 2016). National Institute for Occupational Safety and Health (Putz-Anderson et al., 1997) identified a positive relationship between psychosocial factors and perceived high workload and job dissatisfaction. Devereaux et al. (2002)'s study concluded workers who are significantly exposed to both physical and psychosocial workplace risk factors are more likely to report symptoms of MSDs compared to those who are highly exposed to only one type of risk factor.

Taibi et al. (2021) identified that the risk of MSDs significantly rises with high job demands, job strain, an imbalance between effort and reward, low social support, and perceived unfairness. Furthermore, there is moderate evidence linking low job control, reduced work hours, high levels of workplace bullying, significant hindrance stressors, role conflicts, and interpersonal conflicts to an increased risk of MSDs.

The CCOHS proposes several theories about how stress from psychosocial risks might contribute to the development of MSDs through physiological responses. These theories suggest that psychosocial stress can lead to increased muscle tension, which, in turn puts excessive pressure on joints and soft tissues. Other potential factors include heightened blood pressure and fluid retention, again adding excessive pressure to the joints and soft tissues. There is also the risk of individuals applying excessive force due to a reduced sensitivity to pain resulting from prolonged stress.

Addressing Psychosocial Risks

Although organizations have been focusing on mental health wellness in the workplace by implementing mitigations, especially through educational programs, the COVID-19 pandemic has heightened mental health stress concerns, thereby increasing organizational emphasis on this issue. When many employees worked extensively in isolation, stressed over finances in an uncertain economy, or struggled with juggling basic health and safety for themselves and their families, the collective mindset of the population was significantly affected during this period of time. And this toll translated into increases in the occurrence of documented mental health issues and crises, as evident with the American Psychological Association finding, as part of their 2021 Work and Well-being Survey (Abramson, 2022), that almost 80% of U.S. workers had signs of burnout and the WHO estimated almost one billion (WHO, 2020) people globally as having a mental-health or substance-use disorder exacerbated by the pandemic.

With the increased focus on behavioral health, organizations have taken actions to implement solutions to manage mental health concerns, such as company-funded programs that emphasize counseling and support through Employee Assistance Programs, or programs and policies that promote more balanced life-work practices, or general communications on mindfulness, both at work and at home. However, to address the deeper issues surrounding all the various aspects of psychosocial risks, more needed to be done, particularly around a better understanding of the development of stress and effective means of preventing and controlling them. To confound the issue and add to the complexities, there are various theories surrounding work-related stress which can complicate directions for strategies. These theories include, among many others, Demand-Control, Effort-Reward Imbalance, Organizational Justice, and Job Demand-Resources (Cambridge Companion of Organizational Stress and Wellbeing (Siegrist, 2023). Each theory varies in its proposed mechanisms for the development of stress. However, these theories act under the common expectation that psychological processes: Emotional, cognitive, and behavioral responses in addition to physiological responses to said psychological processes (increased heart rate, muscle tension), are central in understanding how stress develops and how interventions can be developed to prevent, control, or at the very least manage the consequential effects of it.

Given that the *perception of work* serves as an important element in how stress develops, it is not a huge leap to see how work designed to exclude—no matter if unintentional—can have a significant impact on those experiencing feelings of exclusion, be it physical or non-physical. And since psychosocial risk factors are heavily influenced by how workers perceive their work, it is critical that we as ergonomics practitioners maintain a focus on both physical risks *and* psychosocial risks within our ergonomics processes. Without proper measures to manage these factors, such as reducing job demands by fitting the task to the person, establishing support systems, creating inclusive and accessible workspaces, and promoting open communication and engagement, the psychological safety of workers will be greatly compromised, potentially leading to increased risks.

PSYCHOLOGICAL SAFETY

In the upcoming chapter discussing designing for gender diversity, you will read a story about how the safety culture became a fundamental root for injuries due to the lack of fully addressing cultural differences and expectations. You will read how the issue went beyond just the lack of availability of appropriately sized equipment; it reflected a broader oversight in safety practices that didn't fully account for said cultural expectations. You will also learn about how women in high-stress and dangerous work environments often feel internal pressure to stay silent or "make do" with challenges caused by system or equipment design, simply to avoid being perceived as "weak" or a "nuisance." This overcompensation is largely driven by stereotypes, many of which are reinforced *by* the very system design barriers they face. While not intentional, such oversights are common.

Cultural expectations can significantly influence the effectiveness of safety and ergonomics interventions. For instance, in some Asian cultures, respect for authority might discourage individuals from voicing concerns or perceived complaints, which can affect the reporting and addressing of safety issues.

In their 2018 Professional Safety article "Safety Across Cultures: Understanding The Challenges," authors Michael Flynn, Elizabeth Castellanos, and Augusto Flores-Andrade emphasized the importance for safety leaders to recognize and manage cultural differences within the workplace. They highlight the need for leaders to be aware of their own cultural perspectives and understand the power dynamics at play, including their own social position and that of those they manage or work with.

To effectively address these cultural differences, safety leaders must be observant, ask pertinent questions, and maintain an openness to having their own cultural biases challenged. Embracing this openness not only helps in navigating cultural differences but also fosters a psychologically safe environment where all individuals feel valued and heard.

The concept of "psychological safety" has garnered significant attention recently, but it was first introduced back in 1965 by Edgar Schein and Warren Bennis. This term describes an individual's perception of potential threats or rewards when taking interpersonal risks in the workplace, such as sharing thoughts, viewpoints, or ideas. Harvard Business School professor Amy Edmondson, a leading expert in this field, further explains in her book *The Fearless Organization: Creating Psychological Safety in the Workplace for Learning, Innovation, and Growth* (Edmondson, 2018), it as conscientious efforts by leaders within organizations to intentionally cultivate a space where employees feel comfortable sharing information, taking risks, and openly expressing concerns. Doing this creates a culture of innovation and high-performing teams.

In a psychologically safe environment, individuals feel secure in expressing themselves without fearing ridicule from management or colleagues. Conversely, if employees perceive that their ideas might be met with criticism or rejection, the perceived risks outweigh the rewards, leading to a less safe environment. Psychological safety, while related to trust, is distinct in that it specifically addresses the dynamics within a group or workplace setting (Capezio, 2024). Research has shown that a positive psychological safety environment can significantly enhance individual attitudes, team dynamics, and overall worker performance (Capezio, 2024).

Psychosocial Risks, Psychological Safety, Ergonomics, and DEI

Psychosocial risk and psychological safety are deeply interconnected, with each influencing the other. High, uncontrolled psychosocial risks can significantly undermine efforts to create a psychologically safe workplace. Conversely, a lack of psychological safety can exacerbate psychosocial risks by increasing stress and dissatisfaction among employees, thereby creating a negative feedback loop where poor psychological safety contributes to higher psychosocial risks, and vice versa.

Psychosocial risks like exclusion or discrimination can significantly impact psychological safety, making it essential for us to identify and manage these risks effectively. To foster a psychologically safe workplace, there need to be supportive and inclusive environments with open communication, where psychosocial risks are addressed proactively. Our role in this, as ergonomics practitioners, will involve conducting thorough risk analyses to pinpoint potential psychosocial hazards, implementing targeted interventions to mitigate these risks, and designing workspaces and practices that promote inclusion and engagement. In short, as Rachael advocates and what I feel is the most essential objective for our profession, we need to prioritize having an inclusive mindset.

Through anecdotes, historical insights, and thoughtful interviews with colleagues who have shared real-life examples and practical advice, the following chapters in Part II aim to delve deeper into how we, as a profession, can foster an inclusive mindset that embraces the diverse characteristics of today's workforce.

REFERENCES

Abramson, A. (2022, January 1). Confronting burnout: Overcoming stress to reclaim your passion. *Monitor on Psychology*. https://www.apa.org/monitor/2022/01/special-burnout-stress

Capezio, A., Barends, E., Rousseau, D., & Wietrak, E. (2024). Psychological safety: An evidence review. *Scientific summary*. Chartered Institute of Personnel and Development.

Devereux, J. J., Vlachonikolis, I. G., & Buckle, P. W. (2002). Epidemiological study to investigate potential interaction between physical and psychosocial factors at work that may increase the risk of symptoms of musculoskeletal disorder of the neck and upper limb. *Occupational and Environmental Medicine*, *59*(4), 269–277. https://doi.org/10.1136/oem.59.4.269

Edmondson, A. C. (2018). *The fearless organization: Creating psychological safety in the workplace for learning, innovation, and growth*. Wiley.

European Agency for Safety and Health at Work. (n.d.). Psychosocial risks and mental health at work. https://osha.europa.eu/en/themes/psychosocial-risks-and-mental-health

Ferreras, J. (2016, September 1). Depression, anxiety cost Canadian economy almost $50 billion per year: Study. *Huffington Post Canada*. https://www.huffpost.com/archive/ca/entry/depression-anxiety-cost-canadian-economy-almost-50-billion-per_n_11825770

Flynn, M., Castellanos, E., & Flores-Andrade, A. (2018). Safety across cultures: Understanding the challenges. *Professional Safety*, *63*, 28–32.

Karwowski, W., & Zhang, W. (2021). The discipline of human factors and ergonomics. In G. Salvendy (Ed.), *Handbook of human factors and ergonomics* (pp. 1–37). Wiley.
Morgeson, F. P., & Humphrey, S. E. (2008). Job and team design: Toward a more integrative conceptualization of work design. In J. J. Martocchio (Ed.), *Research in personnel and human resources management* (Vol. 27, pp. 39–91). Emerald Group Publishing Limited. https://doi.org/10.1016/S0742-7301(08)27002-7
Oakman, J., Kinsman, N., Macdonald, W., Bartram, T., Keegel, T., & Leka, S. (2016). *Workplace barriers to reducing the incidence of musculoskeletal and mental health disorders* (Research Report No. 106-0716-R02). La Trobe University.
Putz-Anderson, V., Bernard, B. P., Burt, S. E., Cole, L. L., Fairfield-Estill, C., Fine, L. J., et al. (1997). *Musculoskeletal disorders and workplace factors* (Vol. 104). National Institute for Occupational Safety and Health (NIOSH).
Roberts, G. (2019, July 30). Workplace stress is costing European businesses billions. *Entrepreneur United Kingdom.* https://www.entrepreneur.com/en-gb/living/workplace-stress-is-costing-european-businesses-billions/336011
Siegrist, J. (2023). Theories of organizational stress and well-being: A comparative perspective. In L. M. Lapierre & S. C. Cooper (Eds.), *Organizational stress and well-being* (pp. 65–90). Cambridge University Press.
sparks & honey. (2023, October 23). Design, equity and inclusion [LinkedIn post]. LinkedIn. https://www.linkedin.com/posts/sparksandhoney_design-equity-and-inclusion-linkedin-activity-7122599371002494976-FQXF?utm_source=share&utm_medium=member_desktop
Taibi, Y., Metzler, Y. A., Bellingrath, S., & Müller, A. (2021). A systematic overview on the risk effects of psychosocial work characteristics on musculoskeletal disorders, absenteeism, and workplace accidents. *Applied Ergonomics, 95*, 103434. https://doi.org/10.1016/j.apergo.2021.103434
World Health Organization. (2020, August 27). *World Mental Health Day: An opportunity to kick-start a massive scale-up in investment in mental health.* World Health Organization. https://www.who.int/news/item/27-08-2020-world-mental-health-day-an-opportunity-to-kick-start-a-massive-scale-up-in-investment-in-mental-health

6 Ergonomics and the Future of Work

The need to seek a balance between humans and systems has always been part of our ergonomics DNA. Just look at history: From the primitive ax perfectly molded to fit our hands to Hippocrates' thoughtful arrangement of surgical tools for ease of use, the principles of ergonomics were around long before the Industrial Revolution. Fast forward to recent times, and we have seen quite a few shifts that have put this balance to the test. The Industrial Revolution kicked off an era where the interaction between workers/end users and their jobs, tools, and products started to shape ergonomics into the discipline we see today. However, the discipline continues to be shaped by the world around us, expanding the scope of our efforts as ergonomics practitioners. For example, with the ongoing changes brought on initially and continually by Industry 4.0/Industry 5.0, Society 5.0 and "Factories of the Future" initiatives, we are looking at an entirely new work landscape—workers and the environment—that seems to be changing day by day. Changes such as advanced robotics and artificial intelligence (AI) have become staples in the workplace, transforming the way we think about ergonomics, especially as it relates to how workers function and engage in a cyber-physical landscape.

And then there's the COVID-19 pandemic, which accelerated remote work and flex work, changing how we think about ergonomics in home offices and alternative work environments. Suddenly, our dining tables became makeshift desks, stacks of books were repurposed as makeshift risers to position laptop cameras at the right height and avoid the dreaded "Zoom neck." Meanwhile, people scrambled to find "ergonomic" chairs that were both affordable and functional (who would have thought that finding the right chair could feel like a quest for the Holy Grail?).

As workspaces continue to evolve, the challenge for ergonomics practitioners to be effective is continuing to grow, as well. The workforce is getting older and becoming more culturally diverse due to demographic shifts, thus there is a focus on making work and industries more equitable and accessible. The changes in *who* makes up the workforce mean we need to consider even more than ever a wider range of body types, physical and cognitive capabilities, and cultural practices and attitudes. On top of all these changes already impacting our ergonomics scope, there's a significant collective push for sustainability. Society is demanding more from companies, like corporate social responsibility and fair labor practices, leading to a focus on developing eco-friendly products and equitable work processes, regardless of socioeconomic status, while at the same time being safe and operationally efficient.

With all these changes, we're primed for even more transformations that will redefine how we work and interact with our environments. The future of ergonomics is not just about fitting the task to the person; it is about creating inclusive

DOI: 10.1201/9781032658988-9

environments and spaces where the task fits all (ideally), enhancing well-being and allowing people to give more of their authentic selves.

THE CHANGING WORK LANDSCAPE

So, how did we get here? It'll help to look at the (recent) history of work, admittedly, from a Western perspective… and sounding a bit textbookish. Let's start with the First Industrial Revolution, which kicked off in the late 1700s and into the early 1800s. This era was marked by significant innovations, particularly in farming tools and steam power. These advancements transformed agriculture and laid the groundwork for industrial manufacturing. From there, we moved into the Second Industrial Revolution, which took place from the late 1800s to the early 1900s. There we witnessed remarkable progress in automation technology, and this period gave rise to new manufacturing industries, making production processes more efficient and scalable. Then came the Third Industrial Revolution, spanning from the mid-1940s to the early 2000s. This phase was heavily influenced by World War II, leading to significant digital and communications transformations. The development of computers and the internet changed the way we communicate and conduct business.

Now, we find ourselves amid the Fourth Industrial Revolution, also known as Industry 4.0. This current evolution is all about advanced engineering techniques like additive manufacturing and 3D printing, as well as cutting-edge analytics and machine learning. This new era is characterized by the emergence of cyber-physical spaces, where digitalization and connectivity through smart technologies, such as the Industrial Internet of Things, reshape workspaces into adaptive systems that can dynamically change in real-time. We also see a unique integration of humans and machines in these cyber-physical workspaces, working together more closely than ever before, and where the collaboration is not just beneficial but essential (McKinsey & Company, 2021). It is this close collaboration between humans and machines that has brought about Industry 5.0, which is seen as a specific focus within Industry 4.0 that emphasizes the development and use of these advanced technologies in adherence to the welfare of both workers and our planet.

Throughout each phase of industrial evolution, the workforce has also transformed significantly—shifting in terms of who is working, how they are working, and the new skills they acquire as machinery and processes evolve. As a result, the field of ergonomics has expanded its focus to better understand the changing dynamics of work environments and the workers themselves, striving to maintain a balance within these systems.

INDUSTRY 4.0/INDUSTRY 5.0 IN ACTION FOR ERGONOMICS AND DEI

Over the past decade, we have seen significant advancements in tools, technologies, machines, and equipment that enhance our understanding and management of ergonomics within the workplace, redefining what's considered "traditional work settings." As mentioned, with Industry 4.0/Industry 5.0, instead of fully replacing human workers, the emphasis has shifted toward enhancing human-machine collaboration. For example, we now have collaborative robots, or cobots, that work alongside human operators. These robots typically handle repetitive tasks, freeing workers to engage in less fatiguing, yet more complex and rewarding responsibilities (which, I can't help but think Frank and Lillian Gilbreth would be happy with this direction of work given their perspective of striving toward optimal efficiency for more "happiness minutes" for everyone (Stephens, 2024)).

Another promising development was the introduction of passive exoskeletons in assembly lines. These wearable devices assist workers in performing static or awkward tasks, helping to reduce physical strain and prevent injuries. In office environments, innovations such as smart glasses are replacing traditional computer screens, offering a more dynamic and ergonomic experience that enhances collaboration. This synergy between worker and machine not only reduces physical strain on employees but also boosts overall productivity and quality of work. Beyond that, these advancements make tasks that have traditionally excluded people due to high physical or cognitive demands much more accessible than ever.

In terms of risk assessment and analysis activities, we've seen a significant rise in the development and use of tools that leverage digital innovation and AI, revolutionizing the way we conduct workplace risk assessments. We now have quick-access risk assessment apps that collect data or facilitate easy data input, using built-in calculations to automatically evaluate risk factors—quite a shift from the traditional pen-and-paper (and calculator) assessments I cut my teeth on! This advancement allows for faster, real-time evaluations of tasks, and automated reporting systems streamline the process further, enabling organizations to integrate these assessments seamlessly into their workplace management systems.

These innovations not only enhance the speed and accuracy of risk assessment and analysis, but also democratize the ergonomics process, making it more accessible for various stakeholders, including front-line operators, management, and others who may not be ergonomics experts. With the rise of cloud-based platforms, managing extensive data sets, such as risk assessments conducted across a site, is now possible from anywhere, fostering collaboration like never before and increasing employee engagement in the overall process.

As predictive analytics tools have continued to evolve, ergonomics practitioners have been able to quickly and more accurately analyze historical data, including risk assessments, to identify trends and make informed management decisions more efficiently. This data can be seamlessly integrated into other organizational management systems for real-time risk monitoring, helping to pinpoint areas that need prioritization, especially those affecting specific employee groups. This information is crucial for ergonomics practitioners to communicate to management, particularly in relation to corporate strategies focused on Diversity, Equity, and Inclusion, employee engagement, and sustainability.

In training environments, tools like virtual reality (VR) and augmented reality (AR) have gained traction, giving these technologies the ability to vastly improve training processes and facilitate information sharing, which can further help to alleviate cognitive load on workers. Picture a new employee using VR goggles to practice assembly tasks in a simulated setting before they even step onto the actual factory floor. Again, another example of technology that can be used to make previously exclusive tasks more accessible.

As we look to the future, we can expect even more technological advancements, transforming our ergonomics processes and how they can be used to make the workplace safer, more efficient, and more accessible. Advancements such as the development of AI-powered tools will be used to further automate the risk assessment process, particularly during the design phase, to better identify potential hazards and accessibility challenges before they become issues. And with the power of AI, mitigation efforts could be personalized based on individual workers' needs and an analysis of their own historical patterns and behaviors as they interact with their work. And with advancements in tools like VR/AR, communication and training can further evolve to better support diverse workforces, incorporating features such as having features like multi-language real-time translation and personalized training tailored toward an individual's preference, proficiency, and learning styles.

The past tools ushered in with this phase of the industrial revolution, and the advancements we expect to see in the future, not only enhance our abilities to develop, manage, and sustain an effective ergonomics process, but also allow us to better engage a diverse workforce. These tools make the overall ergonomics process more accessible and more easily tailored to the needs of today's diverse workforce. And as this work evolution era continues to advance, it will be incumbent upon us, as ergonomics practitioners, to step up and play a critical role in ensuring that human-centric design remains at the forefront of technological innovation, particularly when it comes to ensuring that the human-machine system is balanced in a workforce that's changing just as rapidly.

THE CHANGING WORKFORCE

Technology is not the only force reshaping the future of human-system interactions in the workplace. Alongside advancements in automation and digitalization, we are experiencing significant changes in the workforce itself. The demographics of the labor population and the nature of the jobs being performed are becoming more diverse than ever. For example, according to the U.S. Bureau of Labor Statistics, while the labor force is expected to increase by over 5% from 2020 to 2030, the number of people 75 and older in the labor force is expected to grow 96.5% in that same time period (U.S. BLS, 2021). This is due to a number of reasons discussed further in Chapter 10, where we delve into designing for an older workforce.

Regardless, this trend presents unique challenges as multiple generations—each with its own work styles, communication preferences, and motivations—work together in the same environment. Younger generations may prefer collaborative, tech-driven approaches, while older workers might value more traditional methods and interpersonal communication. To navigate this evolving landscape, we must find a balance between the physical and cognitive capabilities of an aging workforce and the increasing demands of automation. For example, older employees might benefit from adjustments that accommodate physical limitations, while simultaneously receiving training on new technologies to enhance their productivity.

Additionally, depending on geographical locations, populations are growing more ethnically diverse. This is the case for the U.S. due to factors such as increased immigration, longer life expectancy, and a decline in U.S. birth rates (Wingfield, 2022). While this will result in a workforce that is more representative of various cultural backgrounds, it will, however, necessitate a greater emphasis on cultural competence and sensitivity within organizations to foster effective communication and teamwork, particularly with regards to increased cyber-physical workspaces. For example, where communications are critical to convey important safety messages related to interacting with advanced robotics, collaborative robots (cobots), or automated guided vehicles on the work floor, we will need to ensure that said communications consider potential barriers such as language preferences and proficiency or cultural attitudes that could impact full comprehension and motivation to act upon those communications (more details on designing communications for a diverse workforce are shared in Chapter 12).

Additionally (continuing to use the U.S. as an example of a changing population), demographic trends indicate that the gender balance in the workforce is shifting toward parity. As of the latest reports, women comprise approximately 46.8% of the U.S. labor force (U.S. Department of Labor, 2022) and this number is expected to continue rising, according to projections from the U.S. Bureau of Labor Statistics (U.S. BLS, 2021). This increasing representation of women not only enriches the workplace with varied viewpoints and experiences but also highlights the need for policies and practices that support gender equity (designing for gender diversity is explored in Chapter 7).

THE CHANGE IN HOW WORKERS WORK

In addition to the changing demographics of the workforce, the diversity in the way people work is also evolving significantly. The rise of flexible work arrangements, such as remote work, hot desking, and participation in the gig economy, has transformed the traditional concept of where and how work is performed. This shift was accelerated by the COVID-19 pandemic, which forced many organizations to adapt quickly to remote operations. As such, remote work has become the norm for many employees, leading to a blurring of boundaries between personal and professional life. Ergonomics practitioners now face increased challenges in identifying and mitigating risks associated with these new working conditions. Many workers are taking on multiple jobs or freelance gigs in addition to their primary employment. This can lead to issues like inadequate fatigue management, as individuals may not be aware of how their extended work hours can affect their overall well-being.

To effectively address these challenges, ergonomics practitioners must consider adapting their risk assessments and mitigation strategies. For example, as was evident during the COVID-19 pandemic, ergonomics practitioners increased training and communication on guides for setting up home offices (including the kitchen counter!) and tips on how to incorporate physical activity into their routines while working at home. In addressing the increasing overlap between after-work activities and primary job responsibilities, this could mean implementing educational programs that focus on fatigue management techniques that workers can apply in their daily lives, both at work and outside of it. Additionally, it could mean designing work policies that include flexible scheduling options that allow employees to manage their workloads more effectively, such as staggered work hours or instituting "no meeting" days to help alleviate stress and reduce the risk of burnout. Or it could mean taking a cue from the Industry 4.0 playbook and adopting technological solutions, like wellness apps that track and promote healthy habits.

THE ERGONOMICS PRACTITIONER'S ROLE IN THE CHANGING WORK LANDSCAPE

Now, more than ever, organizations are challenged to manage a work landscape that is evolving at an unprecedented pace. Not only do organizations have to contend with keeping up and staying competitive in an ever-increasing digital world, but they also have to adequately account for changing worker demographics to effectively enhance their ability to be competitive. So, now more than ever, ergonomics and the ergonomics practitioner have a critical role to play.

One such role is our involvement in the design of the cyber-physical workspace. Although there are rapid technological advancements within the workplace, there is, unfortunately, a significant gap between the available skilled talent and the technology required to fully leverage emerging innovations. A 2023 World Economic Forum report predicts a global digital skills shortage by 2030, with an estimated 85 million jobs going unfilled (Trautman, 2023). The report highlights the urgent need for businesses across all sectors to implement long-term strategies to address this gap, focusing on a diverse range of digital skills, providing training through multiple channels, and customizing approaches to meet the specific needs of different industries. In the U.S., one-third of workers lack the necessary digital skills to succeed in today's economy, while Europe struggles with an uneven distribution of skills across countries. Meanwhile, the Asia-Pacific region is projected to see a five-fold increase in demand for digital skills in the near future.

Given this potential gap, we must engage, more than ever, in the initial development and design of these cyber-physical workspaces and processes. This includes being part of the management decisions of when, how, and where to implement technology in the first place, ensuring that the worker is at the center as opposed to a technology-centered approach could see organizations automating for "automation" sake (which could also negatively impact human-machine collaboration, as high automation can potentially reduce vigilance and situational awareness (Endsley & Kaber, 1999).

This should involve us providing guidance and feedback on specific jobs or tasks that could be candidates for consideration for automation or augmentation—such as details on tasks that are assessed as highly repetitive, require high physical effort, have low control, or are highly monotonous, etc. With regard to the design of these collaborative processes and cyber-physical workspaces, we should be providing guidance on the design of systems as related to physical elements such as the design and placement controls, interfaces, etc. as well as cognitive elements such as information processing and communications, etc., with respect to safety, efficiency, and accessibility. Furthermore, we should provide guidance and support for applied research design approaches such as the utilization of adaptive automation like human-in-the-loop. One easy way for ergonomics practitioners to "stay in the loop" themselves is to make sure they are connected and engaged with any group within the organization having these discussions (e.g., Robotics Engineering).

Many of these tasks will require us to be prepared to learn more and do more. This brings us back to the importance of being intentional about diversifying our skill sets. Throughout my career, I have had to constantly learn new tools to "keep up" with all the advancements in technology and research, as well as spending time in other métiers within the organizations I worked at. But I didn't only do it to take advantage of these tools to improve the ergonomics processes and programs I was tasked with developing and maturing; I did it to also take advantage of enhancing my own skill sets for my own career marketability.

So, as technology continues to evolve, it is essential, if not imperative, for ergonomics practitioners to continually seek out opportunities for upskilling. This could include taking certification courses, participating in professional workshops, pursuing academic programs, or attending non-ergonomics industry conferences (I'll save the wild and crazy radio frequency identification (RFID) conference I once attended for another time). Also, don't overlook opportunities like "shadowing" colleagues in other departments, such as those involved in critical management decisions around work organization and automation implementation. This hands-on experience can provide invaluable insights, not just into applying new skills, but into how different parts of the organization connect and impact overall operations. I can say from personal experience that engaging with various areas of the organization greatly enhanced my understanding of its inner workings and put me in front of key decision-makers so that I could advocate for ergonomics. It also contributed to my more holistic approach to ergonomics—an approach that is all but required in this changing work landscape.

PRACTICALLY SPEAKING

This journey to both manage the latest Industrial Revolution and our quest to get a seat at the corporate strategy table won't be without its difficulties. It will demand that we cultivate cultural sensitivity and awareness of the diverse workforces we support, which may require additional efforts on our part, such as training and leadership development to enhance our cultural competence. We'll also need to be more attuned to the physical, cognitive, and psychosocial risks that can emerge from these evolving dynamics—particularly as they relate to a diverse workforce working in cyber-physical spaces. This means understanding the physical composition of our organizations, particularly our workforce, and keeping up with the latest research and advancements in guidelines and tools designed to address specific risks. Additionally, we must become more business-savvy, gaining insight into how our organizations operate, how decisions are made—especially those related to work design—and advocating to be a part of those decisions. This may require some of us—perhaps all of us—to step outside our comfort zones.

While challenges and more work lie ahead for us, we fortunately have a powerful tool in our ergonomics toolbox: A collaborative and participatory approach. As the work environment evolves, from the nature of the workspace to the profiles and demographics of the workers, it is crucial to lean into a culture that encourages open and honest feedback throughout the ergonomics process, especially in the context of Industry 4.0/Industry 5.0. Engagement from a diverse workforce will be vital at every stage—from work design to risk assessments and mitigation strategies to training and communication—because each facet of the human-machine system brings unique risks that necessitate comprehensive involvement to ensure safety and maximize the benefits of new technology.

The future will certainly bring many changes to our field and beyond, many of which we are already beginning to tackle. Yet, throughout this transformation, we must hold onto the principles that have guided our success: designing for prevention, embracing Participatory Ergonomics, and adopting a proactive systems approach to ergonomics programs. By staying alert to emerging trends, learning from past experiences, and adhering to our ergonomics foundation, we will be well-prepared for success in the evolving workplace landscape.

FURTHER READING

- CDC blog post "The Role of Demographics in the Future of Work."
- Forbes article "Understanding The Impact Of Industry 5.0."
- McKinsey & Company article "Hybrid work: Making it fit with your diversity, equity, and inclusion strategy."

REFERENCES

Endsley, M. R., & Kaber, D. B. (1999). Level of automation effects on performance, situation awareness, and workload in a dynamic control task. *Ergonomics, 42*(3), 462–492. https://doi.org/10.1080/001401399185595

McKinsey & Company. (2021, August 23). I, robot: What technology shifts mean for tomorrow's jobs. *McKinsey & Company.* https://www.mckinsey.com/featured-insights/mckinsey-for-kids/i-robot-what-technology-shifts-mean-for-tomorrows-jobs

Stephens, E. (2024). Efficiency and the productive body: The Gilbreths' photographic motion studies of work. *Body & Society, 30*(3), 59–94. https://doi.org/10.1177/1357034X241262176

Trautman, S. (2023, January 16). 5 ways we can develop the digital skills our economy needs. *World Economic Forum.* https://www.weforum.org/stories/2023/01/5-ways-develop-digital-skills-davos2023/

U.S. Bureau of Labor Statistics. (2021, September 8). *Employment and unemployment in the COVID-19 pandemic: An analysis of the U.S. labor market* (PDF). U.S. Bureau of Labor Statistics. https://www.bls.gov/news.release/archives/ecopro_09082021.pdf

U.S. Department of Labor. (2022, February 28). *Working women: A snapshot.* U.S. Department of Labor. https://blog.dol.gov/2022/02/28/working-women-a-snapshot

Wingfield, A. (2022, December 22). U.S. workers are becoming more diverse in race, ethnicity and age—but companies aren't keeping up. *CBS News.* https://www.cbsnews.com/news/workplace-diversity-retention-recruitment-race-ethnicity-age/

7 Designing for Gender Diversity

THE KIDS ARE ALRIGHT

I once had a young design intern approach me, seeking advice on designing truck cabs with women in mind. It was for a project aimed at encouraging more women to join the trucking industry, and my name came up as someone knowledgeable about ergonomics, especially in terms of designing for inclusion.

> **'Gender' & 'Sex' in Ergonomics**
>
> In ergonomics, diversity concerning sex and gender often emphasizes physical differences. According to Johnson et al. (2007), sex is a complex biological construct that includes anatomy, physiology, genes, and hormones, forming a comprehensive human "package." Conversely, they note gender is often conflated with biological sex. In medical discourse and research, gender is seen as separate from sex, further describing gender as the socially prescribed and experienced dimensions of "femaleness" or "maleness" within a society, manifesting at various levels.
>
> Given that physical, social, and psychosocial factors influence how differences are defined because of biological sex and gender identity diversity in the context of the work environment, it becomes challenging to separate the two in design discussions. Therefore, this section will explore how both biological differences (sex) and societal experiences and expectations (gender identity) affect work performance and perception, and how to consider these differences in design. For the purposes of this book, I will refer to the combination of these differences as "gender diversity." A later chapter will delve deeper into design considerations for LGBTQ+ communities, focusing on the psychosocial aspects of gender identity.

Three things struck me from that conversation. For starters, I was thrilled that word had spread about the importance of considering ergonomics when aiming for inclusiveness, even if it was from a primarily physical ergonomics perspective. It felt like a significant milestone in my career—result! Secondly, I was impressed by this intern's awareness that designing for the user is critical for any initiative's success. It reassured me about the future of our profession (the kids are alright!). Lastly, it made me realize that as an "ergonomics expert," I needed to be prepared to provide expertise, even in areas where I might not have immediate answers. The beauty of our profession lies in our ability to find solutions, not necessarily having all the answers at once. In this case, I had to admit that while I had extensive experience in the industry, I hadn't focused specifically on the challenges women face in trucking. However, I saw this as an opportunity to learn and contribute.

So, what did we do? We began by understanding the pain points of female truck drivers. Despite the capability of women who have proven their skill behind the wheel, like Ina Daly from XPO Logistics Freight, the first woman to win a class title at the National Truck Driving Championships (Mulero, 2023), only about 13% of commercial drivers in North America were women according to the Women in

DOI: 10.1201/9781032658988-10

Trucking 2023 Index (WIT, 2023). We learned that the trucking industry's culture and work-life balance often don't align with the needs of many women. There were also specific design issues in truck cabs and trailers, such as handle and step placement for cab access, control accessibility, and limited seat adjustability—all of which disproportionately affected shorter drivers and women.

As an ergonomics practitioner, my role became clear: to address these challenges through thoughtful design adjustments that enhance the overall experience and opportunities for everyone involved in the industry. By optimizing safety, comfort, usability, and accessibility, we not only improve the daily work experience, but we can also contribute to the industry's sustainability.

Throughout this book, I've emphasized that as ergonomics practitioners, we play a crucial role in ensuring that the processes, products, and spaces we design are tailored to meet diverse workers' needs. When these needs aren't met, it can lead to poor user experiences and unintentional exclusion. Our expertise lies in recognizing these needs through our understanding of human anatomy, biomechanics, and cognitive responses to environments. By applying this knowledge effectively, we can design solutions that are not only inclusive but also enhance overall functionality and performance.

DESIGNING FOR SAFETY FOR A GENDER-DIVERSE WORKFORCE

To delve more into this and the role the ergonomics profession can play, I spoke with a very dear colleague of mine, Anuja Patil—an ergonomics and safety professional with experience across a variety of industries, including manufacturing, service, and risk management—on what it means to design for gender diversity, the obstacles and intricacies, and the creative solutions and practices that ergonomics practitioners can use to foster more inclusive design, specifically as it relates to gender diversity.

To Anuja, gender diversity emphasizes the importance of including and representing women in all areas, ergonomics included. This concept advocates for women's adequate representation, respect, and support in work environments, research endeavors, and design methodologies. She notes that it involves catering to women's needs and considerations. While women are generally smaller than men, designing for gender diversity involves more than just accommodating size differences. It is crucial to also consider conditions that disproportionately affect women, such as (Coury et al., 2002; Hooftman et al., 2004; Migliore et al., 2021; NIOSH, 2020; Women in Global Health, 2021):

- Pregnancy.
- Menstrual cycles/menopause.
- The higher prevalence of diseases such as osteoporosis or carpal tunnel syndrome.
- The increased risk of musculoskeletal disorders (MSDs) from movement while bearing loads or with respect to neck/shoulder MSDs in awkward arm postures.
- Increased stress due to low-paying/low status socioeconomic factors.

- Gender-specific religious dress.
- Primary caregiver for children/elderly parents.
- Workplace sexual harassment.

These factors highlight that designing for gender diversity requires a broader approach, addressing not only physical size but also unique health, physiological, and psychosocial challenges, particularly those faced by women who have been traditionally disproportionately impacted within the industry. However, while the list may be shorter, there are gender-specific design concerns that need to be considered from a male perspective, mainly related to health conditions and societal expectations that could disproportionately impact them, such as issues surrounding mental health and stress or over-exertion in the workplace from high physical demand jobs.

In this chapter, we'll explore further the gender diversity considerations we may need to address as ergonomics practitioners, including why it is important to consider them. Topics such as those surrounding personal protective equipment (PPE), work requirements, and stereotypes will be examined, along with how you and your organization can apply these considerations within the design of processes, products, programs, and policies. Doing so will help the organization provide the support system needed for inclusion, ensuring that all are able to be effective and efficient.

PPE

> "The gloves were the problem, they were too big."

This quote comes from a story Anuja shared during our discussion on gender diversity and safety, particularly regarding PPE. It highlights a prevalent issue: PPE often fails to fit individuals properly, especially women, as evidenced by the problem of the "too big" gloves. We'll get back to that story in a bit.

Just to level set, PPE is equipment worn to minimize exposure to physical, chemical, biological, radiological, and other workplace hazards that may result in injury, illness, or even death. This includes items such as safety shoes, hard hats (or helmets), gloves, glasses, face shields, respirators, safety harnesses, and protective clothing, depending on the nature of the job and the specific risks involved. The adequate fit of the required PPE to the worker is critical for ensuring its protective qualities. If PPE does not fit well or suit the individual, they may feel uncomfortable, move less freely, and be less protected from the hazards they are exposed to. However, in many cases, PPE has a "one-size-fits-all" approach and/or is designed based on anthropometric data predominantly derived from male populations (NIOSH, 2020), which can mean that fit can be a challenge with a gender-diverse workforce (and racially diverse workforce as well, to be discussed in Chapter 11). With regards to gender diversity, the lazy "shrink it and pink it" thinking—which is simply making currently available PPE and tools smaller to fit women's hands and adding the color pink to it—was often the modality relied upon when it came to designing (supplying, rather) for women.

Anuja recounted a story to me that many women might find familiar. Early in her career, she worked in a manufacturing role that involved extensive walking on hard surfaces and concrete. Like many women, Anuja had narrow feet and needed specialized safety shoes. Luckily, she had a pair that was "perfect" for her. However, everything changed when she became pregnant. As her feet grew larger and wider, the shoes that once fit well became painfully uncomfortable and worsened the back pain she was also experiencing at the time. Realizing the situation was untenable, she reviewed the company's shoe catalog in hopes of finding a suitable alternative for the duration of her pregnancy. Unfortunately, she found no options that met her needs as, at that time, the range of shoes available to her was very limited and predominantly stylized for men. Anuja joked that finding a pair of shoes designed for women is rare, and finding one that was even remotely stylish was akin to winning the lottery!

Undeterred, Anuja decided to act—she spoke with her manager and the HR department, explaining her situation and asking them to consider providing more inclusive and diverse PPE choices for women. She went so far as to take it upon herself to search for vendors that had more options for safety shoes designed specifically for different foot shapes and sizes and found a vendor that offered shoes that were lighter, softer, and more flexible. Not only was she able to persuade the company to switch to that vendor, she also managed to get them to start a PPE program that would allow PPE to be customized according to worker needs. For Anuja, it was something that stood out as a proud moment for her—proud that she was able to make a positive change not only for her own personal challenges but for others and their diverse needs. She shared,

> "One thing I learned from my previous job in manufacturing was how to advocate for more inclusive and diverse PPE choices for women."

Anuja's story is an example of what many women face on a daily basis. And it shows how something as simple as the improper fit of safety shoes can lead to unnecessary discomfort or, worse, injuries. And we can easily see how this could apply in a variety of PPE situations, not just specific to women, but for other challenges that come along with having a diverse workforce, such as:

- **Tool Belts**: Differences in anthropometry (e.g., hip girth) and weight distribution could be a design concern.
- **Fall-Protection**: Inadequate fit of the harnesses, particularly across the chest and waist areas; potential inaccessibility of key components; chafing or restricted movement.
- **Hard Hats (Helmets)**: Improperly sized hard hats for smaller heads can lead to pain, discomfort, or injury to the cervical spine due to imbalance; could also lead to chafing from padding and straps.
- **Respirators**: Improperly fitted respirators for smaller/thinner faces can lead to inadequate protection (lack of seal) and discomfort.
- **Face Shields**: Improperly sized shields for smaller/thinner faces can lead to inadequate protection and reduced visibility.

- **One-Piece Uniforms**: Can be a challenge for female users when using the bathroom, given that they can be cumbersome to take off; poorly fitting uniforms can lead to discomfort and restricted movement.
- **Passive Exoskeletons**: (which one can argue as a form of PPE)—as with many other PPE worn in traditional male-dominated industries, design of passive exoskeletons are derived from mainly male anthropometric data, which can limit adequate fit and sizing for female users, particularly concerning shoulder width, hip width, hip-to-waist ratios, joint angles, range of motion, and stature. This necessitates activities around anthropometric data standards through the National Institute of Standards and Technology and the American Society for Testing and Materials (ASTM International).

As the world of PPE is vast, so are the examples of challenges when it comes to adequate fit—thus adequate protection—for a diverse workforce. More examples can be found in resources such as the United Kingdom's Trade Union Congress and their "Personal Protective Equipment and Women" (TUC, 2017) guide, Women in Global Health's "Fit for Women" (Women in Global Health, 2021) policy report (targeting PPE for work within the healthcare sector), and the British Standard Institute's Diversity in PPE (Field, 2023) whitepaper.

Additionally, there could also be the *perception* of a company that does not care about the needs of some of its workers. Though I'm sure that was not the intention of the company in Anuja's story, the safety shoe policy could give a *perception* of exclusion.

That brings us back to what we, as ergonomics practitioners, can do to help. We can, as Anuja learned from her experience, be champions and advocates for diverse and inclusive PPE policies and programs. We can and should be involved in the procurement process, especially for tools and equipment sourced for our workers, as our understanding of anthropometry as well as details on the tasks would be valuable guidance to help address safety concerns from improper PPE fit.

Case of the "Too Big Gloves"

Returning to the story of the "too big gloves," Anuja shared an early case from her career as an ergonomist. She was assigned to conduct an ergonomics assessment for a task that had resulted in multiple worker injuries. As Anuja recounted, a female employee she was scheduled to observe missed work because she was suffering from such severe pain in her hand and wrist that she had to visit a doctor for treatment. When she returned to work the next day, Anuja observed her working at the task, exerting a great deal of force and manipulation on a tool. Digging deeper into the cause of the excessive force, Anuja noticed that when she completed the task, the worker appeared to not have a secure grip and that her hand kept slipping. As Anuja continued her observation, she watched the worker continue to struggle with the task. It was at that point Anuja realized that the root cause for the pain the worker was experiencing might be the glove size the female employee was wearing, which was visibly too large. Because the glove was too large, the employee had to exert excessive force to overcompensate for the slippage.

When the employee spoke with Anuja, she confided in Anuja that she had raised the issue of the ill-fitting gloves with the supervisor before, but the gloves that fit her hand properly were extra small and difficult to find. The gloves she needed had to be specially ordered from a supplier and were not available in the vending machines that dispensed the gloves. Although there was a plan for the supervisor to distribute them to employees who required them, sometimes the gloves would run out of stock or the supervisor would forget to place an order.

Anuja also discovered another challenge to the glove PPE program that influenced the availability and use of the right gloves—the employee herself. Taking the time and initiative to speak with the employee, Anuja identified that, due to the employee's culture, she felt hesitant to ask for help or complain about the situation as she didn't want to be perceived as troublesome, or worse yet, lazy, by the supervisor or her co-workers. Anuja devised that the primary problem was attributed to a glove size that did not fit the employee's hand, but there were also other factors that exacerbated the issue, including accessibility to adequate glove sizes and navigating culture within the workplace. To address these challenges, Anuja collaborated with the operations team to adjust the process, ensuring the right-sized gloves were always available for employees who needed them. She also initiated conversations about workplace culture and the new process, using education, communication, and modeling behaviors to ensure employees felt comfortable accessing and using the correct gloves without the fear of judgment.

Anuja's story is an excellent example of how important it is for ergonomics practitioners to not only focus on the technical aspects of an ergonomics assessment, but also on the human and social factors that influence the outcomes. We need to communicate with the workers, understand their needs and challenges, and respect their cultural experiences and perspectives. We also need to ensure that they have easy access to the appropriate PPE and that they feel empowered to use it without worrying about negative consequences from their supervisors or peers. It is also a good example of how the logistical process of obtaining the needed PPE can be a significant factor as well.

ERGONOMICS, DEI AND THE PROCUREMENT PROCESS

When it comes to the logistical details that can impact ergonomics and, by extension, Diversity, Equity, and Inclusion (DEI) efforts, procurement is definitely an area where ergonomics practitioners need to make significant inroads. Organizational procurement activities designed to optimize the acquisition of goods and services typically rely on sourcing negotiations and process/delivery times optimized to be cost-effective and sustainable. However, when procurement activities primarily target budget management and process streamlining, the opportunity to miss critical ergonomics-related needs can occur, especially if those making sourcing decisions are not fully aware of those needs.

It is not uncommon for a single vendor or limited vendors to be sought out to manage all sourcing of safety needs. For example, many companies utilize large distribution industrial suppliers to provide many items, including PPE. In many cases, this works, especially for non-PPE items. However, for PPE, sometimes this sourcing decision can mean a limited selection in variety, fit, quality, or suitability for different workers. When fit is critical, such as with gloves, respirators, glasses, and hard hats, more effort into what is available to properly fit a diverse population is warranted. This includes investigating if the required PPE items are available and tailored for women's needs. As for ergonomics practitioners, it could mean our role in this process is educating our procurement teams about the importance of having the right PPE for a diverse population. We need to be able to explain the reasoning for specific brands, ratings, and sizes for certain PPE items and provide selection guidance based on the science and standards behind the reasoning. We need to be able to provide them with the business case for having more diverse PPE options, such as the potential avoidance of injuries (and the cost associated), increased compliance, and the improvement of employee morale.

One particular roadblock might be that certain types of PPE can be perceived as either prohibitively expensive or challenging to source, depending on their style and specific requirements. When faced with such issues, as Anuja was in her story, it may be necessary to conduct a more thorough risk evaluation. This process helps to build a solid business case for investing in the PPE or to identify alternative operational solutions that could eliminate the need for it altogether. By addressing these concerns, we can ensure that safety measures are both effective and feasible within the constraints of budget and availability.

The U.S.'s General Service Administration (GSA) Section 508 (GSA, 2023) offers guidance for organizations with federal contracts on how to incorporate accessibility into their procurement processes. While the focus is primarily on IT products and services, this guidance can also help organizations, especially those dealing with PPE, to create a more inclusive procurement framework. We've already highlighted the benefits of an accessible procurement process, but the GSA emphasizes several key points to gain executive support:

- Increased productivity through enhanced performance for employees with disabilities and streamlined usability for all employees.
- Choosing accessible products upfront could prevent the high costs of retrofitting solutions later.
- Providing accessible technology could help companies prevent legal fines and lawsuits related to accessibility compliance (where applicable).
- Investing in accessible technology supports an aging workforce and reduces turnover costs by helping employees adapt to changing needs.
- Inclusive technology practices draw in talented employees with disabilities, fostering a diverse and innovative workforce.

The GSA also recommends that organizations assess the diverse needs of their target users, cautioning against relying solely on the presumed "average user." Here, ergonomics practitioners can play a key role by documenting worker demographics, identifying current barriers through task risk assessments, and collaborating with internal experts to establish technical specifications and performance criteria. This is an opportunity to integrate anthropometric, physical, cognitive, and psychosocial factors relevant to the workforce.

For example, in addressing glove specifications (sticking with the glove PPE motif), it is crucial to ensure that size ranges include options for smaller hands and that performance metrics cover material properties, grip, dexterity, and usability aspects like ease of putting on and taking off the gloves.

As potential vendors emerge during the procurement process, ergonomics practitioners can assist in evaluating proposals, reviewing product specifications against the established criteria, and testing samples. During implementation, we can help create training materials, assist with installation and pilot testing, monitor usage, identify any risks, and recommend adjustments based on employee feedback.

Program management might involve ergonomics practitioners participating in regular program reviews, incorporating insights from risk assessments, and providing immediate feedback if any issues arise from incident investigations that highlight opportunities for procurement improvement.

We'll delve deeper into this in Part III, but understanding other organizational departments such as the procurement team and their specific key performance indicators and operational strategies—such as sustainability and innovation goals—can greatly aid in finding common ground for comprehensive changes that benefit both employees (through your ergonomics process efforts) and the company's bottom line.

PPE AND WOMEN: THE EVOLUTION

It is not all doom and gloom when it comes to PPE. Anuja shared her view on how PPE has become more inclusive over time, especially with the availability of shoes designed for women's safety. She added,

> About 10 years ago, there were few options for female workers who needed protective footwear, especially in construction and other male-dominated fields. Now, there are many different types of shoes that cater to women's needs, such as lighter weight, better fit, and more style choices.

And it is not just safety shoes. Christy Caputo recently shared in a National Safety Council blog on the evolution of PPE for women (Caputo, 2023) that the call for all PPE designed with the consideration of fitting women has been loud enough to elicit significant changes—well, changes beyond the old "shrink it and pink it" adage so long relied on. I guess hearing the calls for change *and* seeing how big of a market PPE designed for women can be (to the tune of a projected $8.5 billion in 2027 according to (Technavio, 2023) is enough to get things moving in the right direction!

However, Anuja added, this does not mean that all challenges have been overcome. Women may still face barriers or discrimination when it comes to accessing appropriate PPE in some workplaces and industries. This isn't more evident than in the story Andrea Porter, a doctoral candidate at Texas A&M University researching women's armor fit for law enforcement officers, shared with me regarding her research. As Andrea puts it,

> Women in law enforcement are not only fighting crime, they are fighting their own body armor PPE.

Andrea shared that women experience armor discomfort at a rate almost three times that of their male counterparts (Larsen et al., 2018). Furthermore, the Criminal Justice Technology Testing and Evaluation Center reports that more than 60% of female law enforcement officers say that their body armor does not fit properly and causes pain, abrasions, and other discomforts. Due to these issues, female officers often have difficulty completing job demands such as running—so much so that it universally tops the list of women's most problematic tasks to do while in body armor (Coltman et al., 2020, 2021b; Malbon et al., 2020,2022; Ricciardi et al., 2007). Ideally, the armor and the torso form a single system with the armor secured in complete contact with the torso. However, due to poor fit for many female officers, the vest bounces on the body off-cadence to the runner's steps, dynamically changing the center of gravity (Ramstrand, 2016), making it difficult for the body to adjust. To get the image in your mind, imagine running with a loaded backpack on (with both straps, of course—we are ergonomics practitioners, after all!). When running, the backpack will constantly and forcefully bounce on your back, off rhythm from how you're running. You can see how running with the armor bouncing on the chest and breasts can make the activity a particularly difficult and painful task.

Women have long contended with finding support for the chest and breast area, with the widespread use of bras/brassieres as external support in daily activity. Further still, the use of more supportive sports bras is often employed in high-impact activities for additional support. These are indicative of the breast's inadequate anatomical support during physical activity. However, in the case of the armor vest, its weight plus the added weight of attached equipment can result in a force insurmountable for any current support garment.

Andrea shared how Australian researchers have correlated bouncing armor vests to pain, with medium- and large-breasted women more often reporting pain (Coltman et al., 2021a; Coltman et al., 2021b). The vest's repeated frictional force pounding on the delicate support structures of the breast tissues causes breast pain, clinically termed mastalgia (Health Library, 2024), and damages the Cooper ligaments, an intricate network of thin, hair-like ligaments (Coltman et al., 2017) that provide shape and lift to the breast (Cardoso et al., 2015). Whether the wearer uses her arms to hold the vest in place as she runs or uses her arms to run while the vest bounces freely, movement and speed are drastically hindered either way (Coltman et al., 2021b). The option of securing the vest at the waist would provide more stability and disperse the weight, but unfortunately, that space is already occupied by the duty belt with its assorted tools.

Beyond the discomfort, the design of the armor for many officers of a certain build, both female and male, has been found to be inadequate with regards to protection. For starters, the discomfort can lead to incorrect usage of body armor, thus reducing protection as it is not being used as intended. However, even when worn as intended, there can be issues simply due to its design. In 2022, the United States Federal Bureau of Investigation (FBI) conducted a study that found the commonly used body armor, when tested on certain female and certain male molds, was vulnerable to a bullet, or other projectile, ricocheting off the top center of the front armor panel and into the throat area (U.S. HSGAC, 2024).

The severity of the issue has lawmakers listening. Many efforts have been enacted to address this specific concern, namely proposed legislation in the United States, such as the "Female Officers Ballistic Protection Act" (U.S. Senator Jon Ossoff, 2024), which requires all U.S. Federal departments and agencies that employ law enforcement agents or officers to purchase body armor specifically designed to fully protect female officers. The "Department of Homeland Security (DHS) Better Ballistic Body Armor Act" (U.S. HSGAC, 2024) is another proposed piece of legislation that would require all agencies under DHS to provide their female law enforcement officers, and any law enforcement personnel, with improved ballistic body armor to better protect them in the line of duty. Support for legislation such as these has been bolstered by an earlier national defense bill to help ensure that women in the military have quality and proper-fitting gear, including chest plates and other body armor.

While this specific PPE challenge is still ongoing, we can see that there is progress. But it speaks to the serious challenges that can occur when PPE is not designed to fit or otherwise perform as expected with a diverse workforce—in these specific cases, a gender-diverse workforce. These are good examples of why employers should continue to strive for more inclusion and diversity in their PPE policies and practices.

And that includes us, as champions for ergonomics and advocates for the workers, sitting at the (economically procured, I would assume) procurement table.

WOMEN, RUNNING, AND THE EVOLUTION OF SHOE DESIGN

The challenges of PPE designed for (and with) women are not, unfortunately, a solitary issue. For centuries, the concept of women being in industrial work settings, outside of the home and outside of the arts and fashion, was a foreign concept. So, naturally, industrial work needs and designs intended for women weren't a primary concern. But as more and more women entered the workforce and increased their participation, especially in predominantly male-dominated industries, for work (as noted before) and play, the ask—nay, quite frankly the demand—for the design of PPE, tools/equipment, sports apparel, etc., grew louder.

One such area where the cries grew was in the sport of running. I must admit, despite my background as a former track athlete, running isn't exactly my favorite exercise. In fact, I often joke that the only time you will catch me running is if I am being chased! However, I can't deny the staggering impact and growth of the running (as a sport) industry, which equated to $15 billion in 2022 (Andersen, 2023) with over 620 million participants worldwide, approximately 36% of whom are women.

Not long ago, women were discouraged from participating in running altogether. In a BBC article by Olivier Guiberteau (Guiberteau, 2023) featuring Roberta (Bobbi) Gibb, the first woman to run the Boston Marathon, Bobbi recalled how in 1966 she was denied official entry into the race. This sentiment was rooted in a long-standing belief even among official organizations like the International Olympic Committee and World Athletics (formerly the International Association of Athletics Federations, or IAAF), which was very evident after their decision following the 1928 Olympics (English, 2015) to prevent women from competing in competitions greater than 200 meters. This was because it was seen as "too hard" on women because they were "too frail" and "physically incapable." (I am heavy on the quotes because those types of statements were made purely based on what spectators shared, mostly stemming from comments from the press at the time, as a "not a very edifying spectacle" as coined by reporter Knute Rockne for the *Pittsburgh Press** (Rockne, 1928)). Historian Patricia Vertinksy's book "*Eternally Wounded Women: Women, Doctors and Exercise in the Late Nineteenth Century*" (1994) documented these attitudes, which were perpetuated through the medical discourse at the time, shaping the perception of women being physically incapable of strenuous physical activity such as long-distance running. And these attitudes continued to be held by some in the medical field for decades after that fateful 1928 Olympics.

Bobbi Gibb's determination to defy these beliefs led her to sneak into the Boston Marathon, hiding in bushes until the race began, then joining the runners wearing a hoodie and men's running shoes (since women-specific shoes didn't exist then). Despite the discomfort and pain caused by ill-fitting shoes, Bobbi finished the marathon in three hours, 21 minutes and 40 seconds, which was a time faster than two-thirds of the competitors. And women weren't physically capable of running marathons, eh?

This story highlights how misconceptions about women's physical capabilities not only shaped societal attitudes but also influenced product and equipment design. Until the late 1970s, running shoes were primarily designed for men, with women's shoes merely smaller versions of men's models. Runner's World notes (Beverly, 2016) that this began to change in 1978 when New Balance introduced the W320, the first shoe marketed specifically for women, featuring a snugger heel fit and wider forefoot.

As more attention was given to designing shoes tailored to the biomechanics of women's feet—considering factors like anthropometry and stride—the women's running shoe market expanded significantly. By 2030, it is projected to reach a market size of over $5 billion (Andersen, 2023). This shift not only reflects a more inclusive approach to product design but also illustrates the financial wisdom in catering to gender diversity within the industry. As mentioned before in the case of the shift in the design of PPE, financial incentives can often be a powerful motivator in the market, whether one views this as fortunate or not.

Despite the increasing participation of women in running over the decades, the design of shoes and boots specifically for women recently became a priority for the U.S. military. While there was some focus on ensuring armor fit for women (Sisk, 2018), it wasn't until 2018 that it became a significant legislative focus. Fortunately, this growing attention to providing better gear for all servicemen and servicewomen extended to improving the design and fit of women's combat boots (Benson, 2022). Research highlighting that women are more prone to stress fractures from marching and training (Wentz, 2011)—largely due to being issued boots designed for men but in smaller sizes—has driven this change.

Although women have historically had to "make do" with ill-fitting gear, from running marathons in nursing shoes to enduring painful chafing from body armor, it is now clear that more effort is needed to ensure everyone, regardless of gender, has the best possible equipment for the job, including personal protective gear.

OTHER CHALLENGES OF FAILING TO CONSIDER GENDER DIVERSITY

Beyond the physical risks associated with inadequate footwear and other protective gear that servicewomen have admitted to altering to better fit their bodies (O'Connor & Bergengruen, 2019), not considering gender diversity can have wide-reaching psychological impacts in the form of perpetuated negative stereotypes.

As the saying goes, "every Marine is a rifleman," but another unfortunate saying that describes a stereotype has been "women can't shoot." This was a commonly held belief that was pervasive in the Marine Corps when my colleague, Lindsay Hawkins, served. Aside from personal struggles on the rifle range, Lindsay faced hard facts when she served on the staff of the 4th Recruit Training Battalion at the Recruit Depot on Parris Island from 2004 to 2007. At that time, the 4th Battalion was the only all-female recruit training unit. Each cycle, Lindsay and her peers could see how their female recruits scored compared to their male counterparts on everything from swim qualifications to drill competitions, physical fitness to rifle range scores. It was the rifle range scores that were particularly challenging for Lindsay because it perpetuated the "women can't shoot" stereotype that she knew was unfair.

Historically, this stereotype was reinforced by data showing female recruits had lower qualification averages, 67–78%, compared to their male counterparts, who averaged 85–93% (Germano, 2016). There were anecdotes that the male recruits might have had more experience hunting or playing video games than the female recruits, so perhaps they felt more comfortable around weapons. The counterargument is that a good instructor can coach someone who is eager to learn and does not have bad habits. Something else was in play. Lindsay's insights and a gender disparity investigation by Kater Germano in *Task & Purpose* reveal that the issue might not be women's shooting ability, but rather the lack of gender diversity in the design of rifle qualification standards.

The Marine Corps' Annual Rifle Qualification (ARQ) is meant to gauge a Marine's rifle proficiency through various shooting postures (sitting, prone, and standing) and varying distances (previously from 200 to 500 yards). Lindsay explained that while the postures were intended to provide stability, they, in her opinion, often didn't accommodate the physical differences among Marines. She found that in order to hold the rifle securely against her shoulder in the right position, she wasn't always stable.

> When in the cross-legged seated position, we were told to use particular points on our knees as support. But that wasn't possible for me. In order to hold the rifle correctly I had to raise my knees in, and that wasn't exactly a stable posture!
>
> (Lindsay shared this while demonstrating how she would teeter in the position).

Lindsay also highlighted that the qualification was particularly challenging for shorter recruits.

> It was a problem for anyone shorter than 6 feet... many female recruits are shorter than that.

Her observations were further validated by a female recruit who exceeded 6 feet (about 1830 mm) and set local ARQ records, further indicating to Lindsay that the ARQ was designed with a >6-foot male in mind, making it possibly the de facto standard. This misalignment, Lindsay feels, contributed to the stereotype that women were less proficient marksmen, reinforcing a self-fulfilling prophecy of lowered expectations for female recruits, as Germano also pointed out.

Another challenge Lindsay highlighted regarding the "women can't shoot" stereotype was the pressure to overcompensate when facing other issues, as she didn't want to be perceived as complaining or being a "problem." Lindsay, being left-eye dominant yet right-hand dominant (cross-dominance), found adjusting her shooting stance so the butt of the rifle rested against her left shoulder was the best method for her to account for the cross-dominance. Since she is right-handed, this adjustment felt awkward. Additionally, due to the standard rifle design, the casing ejection port ended up directing spent casings right into her because of the altered posture. While this is not solely a gender diversity issue (as it affects men too, although studies have shown it to be more prominent in women (Dargent-Paré, 1992)), Lindsay felt the added pressure of adhering to social expectations and avoiding complaints, compounded by the stereotype, which created an extra layer of stress for her and others.

The Marine Corps began making several key changes throughout the latter part of the 2010s. These included increasing the representation of women in recruiting materials, enhancing gender integration in boot camps, and restructuring rifle training (Venable, 2020). Notably, the ARQ standards were revised around the same time, better accommodated different (MilitarySpot, 2021) body types and all but eliminating the stereotype that "women can't shoot." One significant change was allowing recruits to rest their weapons on support structures—an adjustment that Lindsay had highlighted as a crucial challenge during her qualification rounds during her time of service. These efforts reflect a broader initiative within the Marine Corps to challenge outdated biases and ensure that all recruits, regardless of gender, have an equitable opportunity to demonstrate their proficiency.

What can happen when gender diversity is not taken into consideration?

It is not a huge leap of the imagination to see how, when gender is not a design consideration—be it in the design of processes, training programs, or policies, many are at risk of being excluded. This exclusion can have several dire consequences—increased safety risks, decreased operational efficiency, and the perpetuation of harmful stereotypes.

Anuja summarizes that by incorporating considerations of gender diversity into the design and implementation of workplace guidelines, processes, and activities, the ergonomics profession is helping to ensure equitable treatment and opportunities for women. Throughout this book, I have, and will, speak a great deal about how critical the user is in user-centered design (the clue is in the name). But it doesn't hurt to speak about it more, because it is *that* critical. Without the user, we can't possibly understand the user's needs, preferences or challenges to a potential design, let alone what pain points we are trying to address. Anything we propose to develop for the intended use of others may look good on paper but could, in actuality, fall short when in action. By involving users throughout the design process, we can gain their insights and ensure that whatever we develop is not only functional but also actually useful to those who will interact with it, including their overall experience, which *will* be shaped by their diverse perspectives and life experiences.

When we are intentional with prioritizing the inclusion of women through our actions and advocacy, we are indeed, as Anuja eloquently puts it, "*recognizing and valuing the unique perspectives and contributions of all...*"

DESIGNING FOR GENDER DIVERSITY = REDUCED STRESS ON ALL

While the previous discussions and examples have been heavily women-centric in terms of exclusive design—and for good reason as described earlier relating to the fact that most product designs originated or were designed specifically for men and the observed inequitable employment and working conditions that have led to the formation of dedicated technical groups to address—inclusive design is essential for everyone.

For instance, the design of mental health services, including communication strategies and visual elements, often targets women, even though men are equally vulnerable to mental health issues. According to the National Alliance on Mental Illness, men die by suicide at four times the rate of women and have higher rates of Attention-Deficit/Hyperactivity Disorder, yet they are significantly less likely to seek mental health treatment (NAMI, n.d.). This disparity can partly be attributed to how mental health services are communicated and marketed. Many mental health campaigns do not resonate with men, which can deter them from engaging with these resources. Additionally, societal stereotypes reinforce the idea that men should be resilient, driven, and self-sufficient, further discouraging them from seeking help when needed. To effectively support everyone, it is crucial to create mental health communications that resonate with all genders, addressing their unique experiences and encouraging open dialogue about mental well-being.

In the workplace, the impact of exclusionary gender diversity design can be seen to impact male counterparts in other ways, such as them taking on more work to accommodate poor process design. From an injury perspective, studies have shown that men have a higher risk of MSDs, specifically related to lifting tasks and back pain, as well as for the relation between hand-arm vibration and neck-shoulder complaints (Hooftman et al., 2004). To delve more into designing for gender diversity from an accommodation perspective in the workplace, I spoke with Ann Hall, a Certified Professional Ergonomist with 20 years of experience and a sought-after leader in this space, having presented on the topic at numerous conferences, such as the Applied Ergonomics Conference and National Ergo Expo, as well as safety-focused events such as the Chicagoland Safety Health and Environment Conference and the Indiana Safety and Health Conference. Ann believes that failing to make the necessary accommodations in the workplace can lead to significant stress, tension, and resentment.

She recounted a case involving a male warehouse employee in his mid-forties who had previously filed grievances. His complaints stemmed from the company's practice of hiring female workers for the same role as his but with exemptions from lifting heavier boxes. He felt it was unfair that the women were not required to handle the same physical demands, and he resented being asked to take on the additional workload.

During a discussion with this employee, Ann realized the potential for such situations to create considerable workplace stress and tension. Some employees might feel resentful, while others could become insecure. This insecurity might drive them to exceed their physical limits in an attempt to prove their value and gain approval from peers. She proposed a common-sense solution:

> Instead of forcing some employees to work in high injury risk situations, what if we asked the company to redesign the job to be more inclusive? This is something we see in many work environments. Jobs have seemingly been designed for the young, healthy male and that doesn't represent the working population anymore. This narrows down the job pool, and thus creates a problem in filling positions. It also prevents companies from having a diverse and equitable environment.

She explained that while efforts to promote diversity are well-intentioned, they can fall short if they do not address the root causes of performance disparities, such as high physical demands. By putting effort into redesigning the job so that it accommodates at least 75% of women (in terms of strength) and others with lower strength thresholds, the job pool can be expanded to include a broader range of candidates. This adjustment also reduces physical and mental stress for all employees, including younger males who might otherwise face undue strain.

Ann also stressed the importance of incorporating DEI perspectives into pre-hire activities, such as functional testing. I had the opportunity to speak with James Rethaber, Vice President of Technical Operations at Fit For Work and an expert in functional testing. He explained that functional testing assesses whether job candidates can meet the physical requirements of a job. There are two stages: Pre-offer, which evaluates physical abilities without medical assessments, and post-offer, which includes medical evaluations after a conditional job offer is made.

He explained that organizations typically implement functional testing for roles that demand heavy physical labor, such as jobs involving lifting, pushing, or highly repetitive motions. He has seen it utilized as a strategic measure to combat high turnover rates and high injury costs in physically demanding positions, particularly in industries with high rates of musculoskeletal injuries, especially among new hires. However, organizations must comply with local laws and regulations, such as the Americans with Disabilities Act and the Equal Employment Opportunity Commission Uniform Guidelines in the U.S., to avoid discrimination. For instance, testing must be job-specific and relevant to the position, and the information collected during testing must align with the appropriate employment stage—whether pre-offer or post-offer. Regular reviews of testing programs are essential to avoid unintended discrimination against certain groups, such as older workers or women.

Ann pointed out that while many companies use these "fit for work" tests to ensure candidates can perform the job safely, ergonomics practitioners still need to address critical questions before and during the implementation of such programs, including questions from a DEI perspective:

- Are the tasks reasonable?
- Are they essential?
- Will the tasks exclude certain populations (e.g., women)? If so, what measures can be implemented to address this issue before testing?

> Many professionals question the ethics of screening employees on work that they probably should not be doing in the first place. How many employees are pushing through their own personal limitations for a chance to have the job they desire when they really should not?

Aside from the decision on whether or not to implement such testing, Ann added, ergonomics practitioners must be prepared to answer these questions to ensure a fair and inclusive workplace.

In my conversation with James about pre-hire functional testing, he noted that the testing process itself can be a useful tool to advance corporate DEI strategies when used to glean valuable insights through data analysis. This information, gathered given the focus on task-specific demands, could be used by ergonomics practitioners to identify areas that need improvement or tasks that should be prioritized for optimization—whether for safety or operational efficiency. For starters, the Physical Demands Analyses (PDAs), which are detailed assessments of the physical, cognitive, and environmental demands of job tasks used as part of the process of developing testing protocols, are extremely valuable tools for the organization. Not only do they provide scientific rigor to develop job-specific tasks utilized in functional testing protocol design (ensuring the protocols are legally defensible), but the information also collected in the PDAs can complement ergonomics, safety, operations, and other assessments to identify tasks that should be prioritized for mitigation—especially tasks that would necessitate the need to even consider functional testing.

He shared a story of how, during a PDA of an industrial bakery mixer, it was found that workers had to lift 25 kg bags from the floor to overhead and place them at waist height for an entire 8-hour shift, which raised the risk of lower back and shoulder injuries. He noted that due to the PDA, a vacuum lift assist device was recommended and implemented, reducing the heaviest item a mixer must lift to just 9 kg. He added,

> This significantly decreases the potential for injury, as well as increases the number of candidates physically capable of performing this job. This is a win-win for everyone.

As another example, Ann shared a story of her own illustrating how outlining essential functions can reveal gaps. While conducting a job demands analysis at a bottling company, she identified the essential and non-essential tasks, along with the physical and cognitive skills required for the job. For the forklift driving position, she pinpointed two essential tasks, one of which involved changing the propane tank, requiring a lift of approximately 28 kgs at shoulder height. This task, although significant, only made up 3% of a 12-hour shift. Despite being a notable risk for all employees, it was particularly challenging for hiring women for the forklift job, even though the company faced a driver shortage. With the tasks clearly defined, the company was able to prioritize their ergonomics efforts, starting with reassessing this task and modifying it to reduce the risk to an acceptable level.

> Before they modified the task, they hired a woman who had been trying to join the team but gave her an exemption on the lift. When she turned out to be one of their best drivers, it truly supported the reevaluation.

It is important to note that incorporating PDAs into a comprehensive ergonomics risk assessment process—especially for screening—presents significant opportunities for ergonomics practitioners to improve risk assessments, as suggested by Li, Gül, and Al-Hussein, (2019). Therefore, even if functional testing is not implemented, adding a PDA process to your ergonomics strategy should still be considered. However, James noted, if functional testing is implemented, the process can be further used to make work more accessible by helping to address inherent biases. James told of another time when, after working with a company to implement a testing process, he received a call from the employer who was unsure how to proceed with a candidate who had an upper limb impairment (missing one hand). He said that the company was concerned about the candidate's ability to perform an assembly job that required a powered screwdriver. After advising them to evaluate him through the same testing process as other applicants, he said that later that same day they informed him that the candidate successfully passed the test. He goes on to add,

> A month later… the company called me to share that he was excelling in his role and remarked that if they had ten more employees like him, they would dominate the industry!

For James, stories like these highlight the importance of having clear and effective processes in place to ensure that testing, if implemented, truly reflects the essential functions of the job and that these processes are applied consistently. He observed that adhering to these procedures enabled a person from an underrepresented demographic (with respect to the manufacturing sector) to apply for the role, participate in the job-specific assessment alongside other applicants, and ultimately secure the position based on their abilities rather than being judged by their limitations. And by utilizing the process to gain clear insight into the job demands that may exclude individuals, we can better identify the necessary steps to address those barriers and create a more inclusive work environment.

PRACTICALLY SPEAKING

So, what does designing for gender diversity actually look like in practice? As discussed earlier in this chapter, having the right tools and equipment for the job is essential—not only for efficiency but, in some cases, for safety. Proper fit can literally mean the difference between life and death. From an ergonomics practitioner's perspective, designing for gender diversity involves advocating for a tool selection program that accommodates various sizes, strengths, and abilities. For example, ensuring that smaller sizes are available for employees who need them is key. Similarly, uniforms and PPE should come in a variety of sizes, with particular consideration given to gender-related physiological changes like pregnancy. This might include adjusting safety shoe policies or providing footwear that accommodates changes in size during pregnancy. It is important to apply relevant anthropometric data when designing PPE programs and ensure that all teams—such as procurement—are educated about these needs.

Additionally, designing for gender diversity means ensuring that workstations are adjustable to accommodate a wide range of body types and sizes. This includes considering the anthropometric differences between genders that could impact workstation functionality. For example, if workstations require excessive reaching or have controls that are too high, it can lead to unnecessary motions, wasting time and decreasing productivity. Advocating for workspaces and work plans that integrate features to support gender-diverse personal needs is also crucial. This could involve providing nursing rooms, childcare facilities, encouraging a culture where it is acceptable for men to discuss mental health issues, or offering flexible work arrangements to help employees manage both their professional and personal/familial/societal responsibilities.

And it means integrating programs like PDAs used for pre-hire functional testing, into your ergonomics process to identify and address potential gender disparities. By doing so, we can make targeted improvements that reduce unnecessary stress on employees or possibly eliminate the need for testing altogether.

It is also important to at least be aware of functional testing programs, as these are often managed by various departments, such as Human Resources, due to their connection to pre- and post-hire activities. Furthermore, ergonomics practitioners need to stay informed about initiatives from other departments that could impact our ergonomics processes and DEI efforts and be ready to offer guidance when needed. Therefore, it is critical that we are made aware of all initiatives within the organization that intersect with our ergonomics processes. Functional testing is just one example of an initiative that may originate outside of the ergonomics domain but still intersect with it, especially with regards to our providing of guidance in designing for gender diversity (as well as other diversity considerations). Other initiatives worth keeping an eye on include shiftwork/shift scheduling policies, remote and flex work policies, fatigue management, employee retention efforts, feminization initiatives, innovation in PPE/procurement initiatives, wellness campaigns, job accommodations, and Return To Work programs (the latter two will be discussed in more detail in the upcoming chapters).

While these design principles and practical actions are especially important for gender diversity, they are not exclusive to women. *Designing for all by considering all* is a principle that can benefit everyone in the workplace. By using tools and strategies focused on inclusive work design, ergonomics not only addresses physical comfort and safety but also helps create a work environment where all employees, regardless of gender, feel supported and valued. This support spans the entire system: work processes, tools, equipment, PPE, communications, programs, and policies. By intentionally considering and incorporating the gender-diverse needs of employees into our designs and initiatives or expanding our efforts to support their needs, ergonomics can play a key role in promoting a more inclusive and equitable workplace.

FURTHER READING

For additional reading on designing and mitigating challenges related to gender diversity, consider the following:

- *Invisible women: Data bias in a world designed for men* by Caroline Criado Perez (2019)
- *Designing for Sex and Gender Equity* by Isabel Prochner (2022)

REFERENCES

Andersen, J. (2023, September 4). *Running shoes statistics*. RunRepeat. https://runrepeat.com/running-shoes-statistics

Benson, J. (2022, February 3). *DEVCOM leads effort to make warfighter footwear more inclusive*. U.S. Army. https://www.army.mil/article/253700/devcom_leads_effort_to_make_warfighter_footwear_more_inclusive

Beverly, J. (2016, November 18). 50 years of mostly fantastic footwear innovation. *Runner's World*. https://www.runnersworld.com/gear/a20834019/50-years-of-mostly-fantastic-footwear-innovation/

Caputo, C. (2023, October). *Evolution of women's PPE*. National Safety Council. https://www.nsc.org/getmedia/0426ca96-c42c-4341-bbb9-0a8ccca1ff7b/gps-pes-christy-caputo111423.pdf

Cardoso, A., Santos, D., Martins, J., Coelho, G., Barroso, L., & Costa, H. (2015). Breast ligaments: An anatomical study. *European Journal of Plastic Surgery, 38*(2), 91–96. https://doi.org/10.1007/s00238-014-1024-7

Coltman, C. E., Steele, J. R., & McGhee, D. E. (2017). Breast volume is affected by body mass index but not age. *Ergonomics, 60*(11), 1576–1585. https://doi.org/10.1080/00140139.2017.1330968

Coltman, C. E., Steele, J. R., Spratford, W. A., & Molloy, R. H. (2020). Are female soldiers satisfied with the fit and function of body armour? *Applied Ergonomics, 89*, 103197. https://doi.org/10.1016/j.apergo.2020.103197

Coltman, C. E., Brisbine, B. R., & Steele, J. R. (2021a). Bra-body armour integration, breast discomfort and breast injury associated with wearing body armour. *Ergonomics, 64*(12), 1623–1633. https://doi.org/10.1080/00140139.2021.1951849

Coltman, C. E., Brisbine, B. R., Molloy, R. H., Ball, N. B., Spratford, W. A., & Steele, J. R. (2021b). Identifying problems that female soldiers experience with current-issue body armour. *Applied Ergonomics*, *94*, 103384. https://doi.org/10.1016/j.apergo.2021.103384

Coury, H. J., Porcatti, I. A., Alem, M. E. R., & Oishi, J. (2002). Influence of gender on work-related musculoskeletal disorders in repetitive tasks. *International Journal of Industrial Ergonomics*, *29*(1), 33–39. https://doi.org/10.1016/S0169-8141(01)00047-6

Dargent-Paré, C., De Agostini, M., Mesbah, M., & Dellatolas, G. (1992). Foot and eye preferences in adults: Relationship with handedness, sex, and age. *Cortex*, *28*(3), 343–351. https://doi.org/10.1016/S0010-9452(13)80144-X

English, C. (2015, October 8). Not a very edifying spectacle: The controversial women's 800-meter race in the 1928 Olympics. *US Sport History*. https://ussporthistory.com/2015/10/08/not-a-very-edifying-spectacle-the-controversial-womens-800-meter-race-in-the-1928-olympics/

Field, K. (2023). *Diversity in personal protective equipment: A white paper*. BSI Group. https://www.bsigroup.com/globalassets/documents/ppe/diversity-in-ppe-whitepaper.pdf

Germano, K. (2016, March 30). The Marine Corps doesn't have the data to validate the status quo for women. *Task & Purpose*. https://taskandpurpose.com/news/marine-corps-doesnt-data-validate-status-quo-women/

Guiberteau, O. (2023, August 29). How the new World Athletics rules will impact female athletes. *BBC Sport*. https://www.bbc.com/sport/athletics/66615089

Health Library: Evidence-Based Information. (2024, February). *Mastalgia* [Online]. https://search.ebscohost.com/login.aspx?direct=true&db=nup&AN=2009867024&authtype=shib&site=eds-live&scope=site&authtype=shib&custid=s8516548

Hooftman, W. E., van Poppel, M. N., van der Beek, A. J., Bongers, P. M., & van Mechelen, W. (2004). Gender differences in the relations between work-related physical and psychosocial risk factors and musculoskeletal complaints. *Scandinavian Journal of Work, Environment & Health*, *30*(4), 261–278. http://www.jstor.org/stable/40968788

Johnson, J. L., Greaves, L., & Repta, R. (2007). *Better science with sex and gender: A primer for health research*. Women's Health Research Network.

Larsen, L. B., Andersson, E. E., Tranberg, R., & Ramstrand, N. (2018). Multi-site musculoskeletal pain in Swedish police: Associations with discomfort from wearing mandatory equipment and prolonged sitting. *International Archives of Occupational and Environmental Health*, *91*(4), 425–433. https://doi.org/10.1007/s00420-018-1292-9

Li, X., Gül, M., & Al-Hussein, M. (2019). An improved physical demand analysis framework based on ergonomic risk assessment tools for the manufacturing industry. *International Journal of Industrial Ergonomics*, *70*, 58–69. https://doi.org/10.1016/j.ergon.2019.01.004

Malbon, C., Knock, D. C., Critchley, D. R., & Carr, P. D. J. (2020). The effect of breast size and bra type on comfort for UK female police officers wearing body armour. *Applied Ergonomics*, *84*, 103012. https://doi.org/10.1016/j.apergo.2019.103012

Malbon, C., Knock, C., Critchley, R., & Carr, D. J. (2022). The effect of underwired and sports bras on breast shape, key anthropometric dimensions, and body armour comfort. *Police Journal*, *95*(3), 436–458. https://doi.org/10.1177/0032258X211011619

Migliore, M. C., Ricceri, F., Lazzarato, F., & d'Errico, A. (2021). Impact of different work organizational models on gender differences in exposure to psychosocial and ergonomic hazards at work and in mental and physical health. *International Archives of Occupational and Environmental Health*, *94*(8), 1889–1904. https://doi.org/10.1007/s00420-021-01720-z

MilitarySpot.com. (2021, October). *The big change to annual rifle qualification*. MilitarySpot .com. https://www.militaryspot.com/news/the-big-change-to-annual-rifle-qualification

Mulero, E. (2023, April 28). Ina daly's sweet 16 win puts her on road to NTDC again. *Transport Topics*. https://www.ttnews.com/articles/ina-daly-arizona-grand-champ

National Alliance on Mental Illness. (n.d.). *Are there mental health resources and support specifically for men?* NAMI. https://helplinefaqs.nami.org/article/449-are-there-mental-health-resources-and-support-specifically-for-men

National Institute for Occupational Safety and Health (NIOSH). (2020, December 14). Exoskeletons and health equity: A critical look. *NIOSH Science Blog*. https://blogs.cdc .gov/niosh-science-blog/2020/12/14/exoskeletons-health-equity/

O'Connor, E., & Bergengruen, V. (2019, March 8). Women in the military are struggling with health care that doesn't address female-specific issues. *BuzzFeed News*. https://www .buzzfeednews.com/article/emaoconnor/woman-military-doctors-female-problems -health-care

Perez, C. (2019). *Invisible women: Data bias in a world designed for men*. Abrams Press.

Prochner, I. (2022). *Designing for sex and gender equity*. Routledge.

Ramstrand, N., Zugner, R., Larsen, L. B., & Tranberg, R. (2016). Evaluation of load carriage systems used by active duty police officers: Relative effects on walking patterns and perceived comfort. *Applied Ergonomics*, *53*, 36–43. https://doi.org/10.1016/j.apergo .2015.08.007

Ricciardi, R., Deuster, P. A., & Talbot, L. A. (2007). Effects of gender and body adiposity on physiological responses to physical work while wearing body armor. *Military Medicine*, *172*(7), 743–748. https://doi.org/10.7205/milmed.172.7.743

Rockne, K. (1928, August 3). Yankees have another dull day in Olympics. *Pittsburgh Press*, Sports Section, p. 21.

Sisk, R. (2018, May 14). Dunford pushes services to move faster on body armor fitted for women. *Military.com*. https://www.military.com/kitup/daily-news/2018/05/14/dunford -pushes-services-move-faster-body-armor-fitted-women.html

Technavio. (2023, October 16). Personal Protective Equipment (PPE) market size for women to grow by USD 8.35 billion from 2022 to 2027; the market is fragmented due to the presence of prominent companies like 3M Co., Ansell Ltd., and Biffa Plc, and many more. PR Newswire. https://www.prnewswire.com/news-releases/personal-protective-equipment-ppe-market-size-for-women-to-grow-by-usd-8-35-billion-from-2022-to-2027--the-market-is-fragmented-due-to-the-presence-of-prominent-companies-like-3m-co-ansell-ltd-and-biffa-plc-and-many-more---t-301997762.html

Trades Union Congress. (2017). *Personal protective equipment and women: A guide for union reps*. Trades Union Congress. https://www.tuc.org.uk/sites/default/files/ PPEandwomenguidance.pdf

U.S. General Service Administration (GSA) (2023). *Universal design and accessibility*. U.S. General Service Administration Section 508. https://www.section508.gov/develop/universal-design/

U.S. Senate Homeland Security and Governmental Affairs. (2024, May 13). Peters and Britt introduce bipartisan bill to equip female DHS officers with next-generation body armor technology. *U.S. Senate Homeland Security and Governmental Affairs*. https://www .hsgac.senate.gov/media/dems/peters-and-britt-introduce-bipartisan-bill-to-equip -female-dhs-officers-with-next-generation-body-armor-technology/

U.S. Senator Jon Ossoff. (2024, August 2). Sens. Ossoff, Cornyn introduce bipartisan bill to upgrade body armor for female federal law enforcement officers. *U.S. Senator Jon Ossoff*. https://www.ossoff.senate.gov/press-releases/sens-ossoff-cornyn-introduce -bipartisan-bill-to-upgrade-body-armor-for-female-federal-law-enforcement-officers/#:~:text=This%20week,%20Sens.%20Ossoff%20and%20John%20Cornyn

Venable, H. (2020, October). How the few become valued: Improving recruitment and retention of female service members. *Proceedings of the U.S. Naval Institute*. https://www.usni.org/magazines/proceedings/2020/october/how-few-become-valued-improving-recruitment-and-retention-female

Vertinsky, P. (1994). *Eternally wounded women: Women, doctors, and exercise in the late nineteenth century*. University of Toronto Press.

Wentz, L., Liu, P. Y., Haymes, E., & Ilich, J. Z. (2011). Females have a greater incidence of stress fractures than males in both military and athletic populations: A systematic review. *Military Medicine*, *176*(4), 420–430. https://doi.org/10.7205/milmed-d-10-00322

Women in Global Health. (2021, November 1). Fit for women: The gendered impact of COVID-19 on the health workforce (Report). *Women in Global Health*. https://womeningh.org/wp-content/uploads/2022/11/WGH-Fit-for-Women-report-2021.pdf

Women in Trucking Association. (2023, June 28). WIT index 2023 executive summary. *Women In Trucking Association*. https://www.womenintrucking.org/hubfs/WIT%20Index/WIT%20Index%202023%20Executive%20Summary%206.28.23.pdf

8 Designing for Diverse Abilities

Nine-year-old Ashley had the biggest smile. Her smile was the only movement she could control, as she had a form of muscular dystrophy that severely limited her mobility and motions. But when she smiled, she brightened up the entire room.

The story of Ashley was told to me in conversations with Gary Downey, a Professional Engineer and Certified Professional Ergonomist with The Ergonomics Center at NC State University. Gary has had a plethora of experiences in his 25 years, notably having worked for organizations in Kansas as a rehabilitation engineer providing assistive technology services for people with disabilities. Ashley was one such client Gary worked with, using ergonomics to enhance her quality of life.

He tells the story of how they were able to hook her up with a switch with an infrared light that could be pointed at her cheek. And as Ashley consciously moved her cheek, she was able to communicate by selecting words or phrases on a specially designed augmentative device. Gary shared,

> For the first time, Ashley was able to tell her mom "I love you mom." And if the tears weren't flowing on that!... It was like we unlocked this person who was in a shell of a body.

As ergonomics practitioners, we are, for the most part, ultimately in this field to improve quality of life. And what Gary and the team of occupational, physical, and speech therapists he worked with did to aid patients like Ashley is a prime example of how ergonomics is essential to designing for inclusion, particularly for persons with disabilities.

The Importance of Semantics

As professionals advocating for inclusivity, it is crucial to acknowledge not only the importance of what we discuss but also how we engage in these conversations, particularly when referring to individuals from diverse communities. The language we use is more than just a choice of words—it plays a significant role in how we connect with people, build relationships, and drive positive change. Semantics matter, especially when we aim to foster understanding, create meaningful dialogue, and inspire action. Throughout this book, I have chosen to use a mixture of both person-first and identity-first language, as recommended by the U.S. National Institutes of Health. But it is essential to recognize that different people have different preferences regarding how they wish to be addressed. This is a matter of respect for individual autonomy, and the intention is never to create division or dismiss anyone's preferred terms.

For ergonomics practitioners, I offer two key recommendations: First, be mindful that the language we use when communicating, whether within our workplaces or across the broader ergonomics community, is an important aspect of fostering inclusivity. It is a conversation worth having, as our words shape the environment of acceptance we strive to create. Second, much like any initiative aimed at promoting diversity, equity, and inclusion, it is essential to remain open to learning, adapting, and evolving our language when necessary. This includes extending patience, grace, and understanding to others as we navigate these discussions and adjust along the way.

DOI: 10.1201/9781032658988-11

I sat down with Gary to ask him what should be considered, either by those working within the ergonomics space or those managing DEI strategies, when designing specifically for people with physical disabilities.

To Gary, having a physical disability means that someone's body has a certain level of ability that makes it difficult or nearly impossible to perform common tasks that most other people take for granted. That was the case with Ashley. Her muscular dystrophy diagnosis meant that her mobility was impacted to the point that not only was she not able to participate in daily living tasks, but her ability to communicate and engage with the world around her was also severely impeded.

Gary notes that it can be tempting to group physically disabled people into a single group, but that would be like, he adds,

> ...lumping everyone with green eyes into a single group. The physical disability, or eye color, is just one aspect of who they are.

It is important to remember that someone with a disability is first and foremost a person. That disability is simply one facet of their identity, not the whole picture. It is part of what shapes their life experiences and perspectives, which need to be valued and considered in our designs. Valuing the diverse viewpoints of individuals with disabilities can lead to more inclusive and thoughtful designs, as every perspective adds depth and insight, helping us create solutions that work better for everyone. In doing so, we also foster a sense of belonging.

According to the World Health Organization (WHO), as of 2024, an estimated 1.3 billion people have or have experienced a significant disability. When it comes to designing for the inclusiveness of persons with disabilities, we obviously need to recognize that "disability" is a very broad and diverse term. The Centers for Disease Control defines disabilities as "*...any condition of the body or mind (impairment) that makes it more difficult for the person with the condition to do certain activities (activity limitation) and interact with the world around them (participation restrictions).*"

Disabled individuals represent a very diverse population. Not to mention, there are a wide variety of physical and mental disabilities, such as loss or limitations in limb function, vision, hearing, or impairments in learning or thinking. Furthermore, the WHO states that a disability has three dimensions, including impairment in a person's body structure or function, or mental functioning; activity limitation; and participation restrictions in activities of daily living. In addition to the three dimensions of a disability, it can also reflect an impairment at any point in a person's life, such as conditions that are:

- Present at birth and may affect functions later in life.
- Associated with developmental conditions that become apparent during childhood.
- Related to an injury (for example, traumatic brain injury or spinal cord injury).

- Associated with a long-standing condition that can cause a disability such as vision loss, nerve damage, or limb loss (e.g., diabetes).
- Progressive, static, or intermittent.

DESIGNING FOR PEOPLE WITH PHYSICAL DISABILITIES

Disabilities come in many forms, each with its own set of unique needs and challenges. To effectively address these diverse requirements and foster more equitable and inclusive environments, products, and job tasks, Gary suggested two key approaches: accommodations and proactive design.

Accommodations

Gary's first recommendation is to focus on the accommodations process, which is the process of evaluating the unique characteristics and requirements of a person or a group of people, for example, those who use wheelchairs, and modifying the work accordingly to make it accessible and suitable for their abilities. This approach is reactive and involves making adjustments or modifications to existing environments, products, or job tasks to better meet the needs of people with disabilities as they are identified. Gary explains,

> It is a little like the reactive approach in ergonomics. Here we have this existing task that isn't working or is hazardous and we need to modify or change it so someone can do it safely. An accommodation is similar, here we have a house porch with steps; let's add a wooden ramp onto it so that the homeowner who uses a mobility device can go inside or outside easily. The reactive accommodation approach works for many people especially when the solution is tailored to a specific person.

The concept of accommodations, especially in the workplace, emerged mostly from social movements that influenced and shaped governmental policies in the United States and other countries. One example was the disability rights movement, which asked for equal access, inclusion, and participation for people with disabilities in all areas of society, such as education, transportation, housing, and employment. Another example was the movement of veterans returning from wars that aimed to assist them and other people with disabilities to be independent. This movement led to the creation of disability-specific programs and services, such as the Vocational Rehabilitation Act of 1973, which set up the state-federal vocational rehabilitation system in the United States. These movements, including the civil rights movement in the United States during the 60s, helped to enact the Americans with Disabilities Act (ADA) of 1990, which is regarded as the most comprehensive and influential disability rights legislation in history.

Title 1 of the ADA within the U.S. describes accommodations as modifications or adjustments to a job or the work environment. These are changes that employers make to help employees with disabilities or special needs perform their work. Also described by the ADA as "productivity enhancers," they are for the benefit of all workers and not to be seen as "special treatment" for certain individuals. Accommodations can include assistive technology, different work schedules, adapted work environments, or flexible leave policies. These accommodations are to be considered "reasonable," meaning that the accommodation does not impose undue hardship or expense on the employer.

According to Global Disability Rights Now, different countries have different laws and regulations regarding job accommodations for people with disabilities. For example, in some countries, like the U.S., Canada, Australia, and the U.K., there are laws that protect people with disabilities from discrimination at work. These laws require employers to make reasonable accommodations to the job or the workplace to help employees with disabilities do their work. Other countries utilize other measures to ensure employment or support for persons with disabilities in the workforce, such as proportional hiring requirements and/or governmental financial support in the form of grants and tax exemptions. How well these laws and systems are carried out and followed in different countries depends on things like the resources they have, how well they know their rights, and how they view disability in their culture. The main legal basis and right for job accommodations in the United States comes from the ADA. The ADA identifies disability as a condition that affects one or more major life activities significantly, either physically or mentally, or having a history or perception of such a condition, and forbids treating people differently because of disability in employment, and in services, programs, and activities that are public or private. The type and level of the disability, the job requirements, and the needs and abilities of the person can affect what kind of accommodations are needed.

The lack of a precise definition for "reasonable" under the ADA can lead to varying interpretations of what constitutes a reasonable accommodation. This variability depends on factors such as the size and scope of the accommodation and the evolving nature of technology. This ambiguity can create confusion and, in some instances, lead to disputes. Therefore, it is crucial that the accommodation process be collaborative and iterative. Although this process may be time-consuming for the individual requesting the accommodation, it is essential to ensure their input is fully considered so that the accommodation effectively addresses their needs and expectations.

Ann (Hall, who you read about in Chapter 7) also suggests that ergonomics practitioners utilize available resources in their respective countries. For example, in the U.S., the Job Accommodations Network (JAN) offers valuable support. Established by the ADA and managed by the U.S. Department of Labor, JAN provides free and confidential information and consultation on job accommodations. The network assists both employers and individuals with disabilities in identifying and implementing suitable and reasonable accommodations in the workplace. The JAN offers practical examples, general principles, publications, training, and technical assistance on a wide range of topics related to job accommodations and disability employment.

Gary highlighted a prevalent issue with many recent accommodations designed to assist people with disabilities. He noted that these accommodations are often implemented as distinct, specialized features for "the disabled," rather than being integrated as universal elements of the environment. While these solutions may effectively address the needs of individuals, they can also unintentionally create a sense of separation and difference from the broader society. Gary argued that this approach does not foster true inclusion and accessibility, as it can perpetuate stigma and reinforce stereotypes about disability.

Instead, Gary advocated for designing solutions that are both functional and aesthetically integrated into the environment. By creating accommodations that blend seamlessly with their surroundings and are visually appealing, these solutions can be made more acceptable and welcoming to everyone. For example, ramps that are thoughtfully incorporated into the landscape or signs that use a combination of text and symbols not only aid people with disabilities but also benefit other groups such as seniors, children, and tourists. Gary believes that by making these accommodations natural and integral parts of the environment, we can promote a culture of inclusion and respect. This approach helps avoid the perception of accommodations as mere add-ons and supports a more accessible and inclusive world for all.

This philosophy of designing for universal benefit segues into Gary's second recommendation of focusing on proactive design in the form of utilizing Universal Design Principles—a proactive approach that considers the needs of people with disabilities from the start, rather than as an afterthought. (Details about the Universal Design Principles were shared in Chapter 3).

Gary shared that many examples of Universal Design are so simple and elegant that they blend in with the surroundings and benefit everyone. For instance, instead of adding a ramp to the porch of a house, we could design a gently sloped walkway from the driveway to the entrance, with some plants or flowers along the edges. And we could also make sure that the front door has no steps and is wide enough for a wheelchair or scooter to enter easily. These are subtle details that most people would not even notice, but they make a big difference for people with disabilities. He added,

> It is designed in, not added on, which is a much better way to include, welcome, and value people.

Examples of Universal Design used in public spaces shared by Gary can be found in Figures 8.1—8.5 Examples of Universal Design can also be found within the workplace/office as well (Figures 8.6—8.10)

Beyond the differences in opinion around the definition of what constitutes "universal design," there is, as with any new or novel concept, general resistance to change. We all know change can be hard for many, and that's the case with changes to conventional design methods, with some naysayers thinking it is too costly, complicated, and unnecessary.

THE CHALLENGES

There are genuine obstacles to designing for people with physical disabilities, one being resistance to Universal Design. Gary pointed out that many times, there's been a reactive stance when it comes to making things more accessible, which can result in unattractive, makeshift modifications. There are those who do not readily see the

FIGURE 8.1 Image of a man walking through automated sliding doors (source: stockscar/Envato); Universal Design public example of automated doors.

FIGURE 8.2 Image of a man rolling luggage over sidewalk curb cut (source: osbmxhouse/Envato); Universal Design public example of curb cuts.

FIGURE 8.3 Image of a hand opening a door with a lever handle (source: valeriygoncharukphoto/Envato); Universal Design public example of lever handles.

FIGURE 8.4 Image of a person placing hands under automated dispenser (source: StiahailoAnastasiia/Envato); Universal Design public example of automated dispensers.

FIGURE 8.5 Image of a cross-walk signal (source: TonyTheTigersSon/Envato); Universal Design public example of crosswalk.

FIGURE 8.6 Image of a woman standing and working at an adjustable desk (source: Wavebreakmedia/Envato); Universal Design at-work example of adjustable desks.

FIGURE 8.7 Image of a woman working from home (source: seventyfourimages/Envato); Universal Design at-work example of flexible work policies.

FIGURE 8.8 Image of a group of people having a virtual meeting (source: YuriArcursPeopleimages/Envato); Universal Design at-work example of closed captioning/translation subtitles for virtual meetings.

FIGURE 8.9 Image of workers in an office discussing business documents (source: Pressmaster/Envato); Universal Design at-work example of providing communications in multiple mediums.

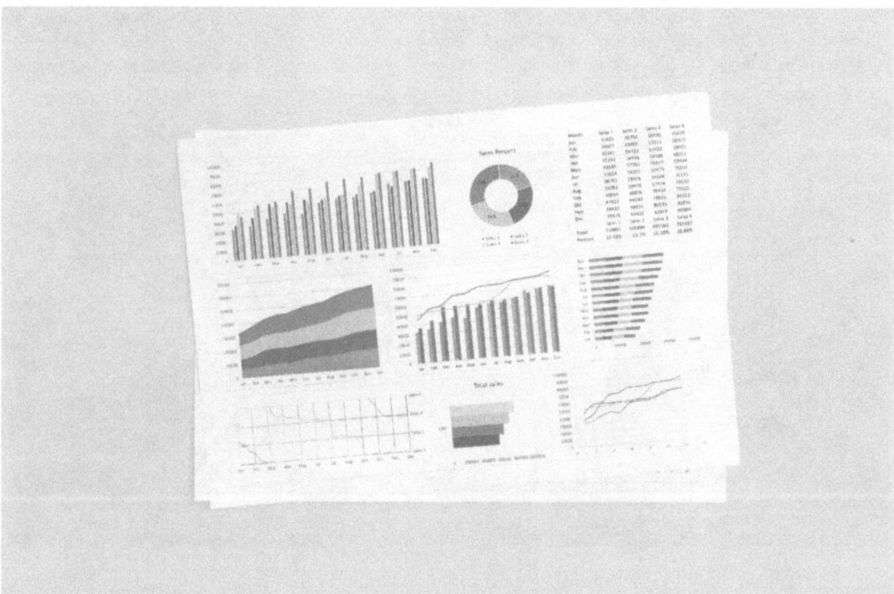

FIGURE 8.10 Image of business graphs and charts on table (source: LunaKate/Envato); Universal Design at-work example of providing easy-to-read data graphs/charts.

benefits of designing for a diverse group of end users, particularly if, in their opinion, that group is such a small minority, and there is the potential of reduction in quality and performance of their designs. Unfortunately, a lot of this thinking is driven by inherent biases and, unfortunately still, prejudices against marginalized communities. It is that lack of empathy and understanding (or willingness to understand) that fuels this challenge to Universal Design.

Another significant challenge impeding the effective design for persons with disabilities is the potential of an ineffective accommodations process within the organization. This can be due to a variety of factors (Siler, 2018), making it difficult for employees to receive the support they need. One major issue is the lack of a clear, accessible process for workers to follow. If employees are unaware of how to request accommodations or what the process entails, they may not even attempt to ask for help, leading to unmet needs. This lack of awareness, coupled with poor communication about the availability of accommodations, can render the entire process ineffective. Additionally, the complexity of individual needs often requires a more personalized approach. A one-size-fits-all, or "cookie-cutter," solution may not address the unique challenges faced by different employees. When the accommodations process is too rigid and doesn't allow for flexibility or customization, it can fall short of effectively meeting diverse requirements. This is particularly true when employees' needs are varied, requiring solutions that can evolve over time as circumstances change. Due to this, the organization might not have the dedicated resources or expertise needed to manage these accommodations effectively. Without trained personnel who are knowledgeable about the specific types of accommodations or legal requirements, the process may falter. In some cases, organizations may lack the capacity to provide the necessary tools or adjust workplace environments in a way that genuinely supports all employees. While this is seen as more of an issue with cognitive disabilities, as they can be less obvious than physical disabilities and thus less likely to be seen as a need, the challenge remains for anyone seeking accommodation.

As ergonomics practitioners, we see a significant opportunity for us to make a real difference in this area. If the accommodations process isn't already part of the ergonomics function within the organization, it is crucial that we engage with it more closely (see the conclusion for more thoughts on how we can contribute to this important area).

Engage

Regardless of whether it is due to resistance to adopting Universal Design Principles, the implementation of an accommodations process that is lacking in effectiveness, or the general lack of knowledge in designing for physical disabilities, these challenges are hindering organizations' ability to be inclusive. When I asked Gary about these challenges, he shared that a proactive approach is essential. To Gary, many people do not understand what a person with a disability requires or how to make them feel comfortable and accepted in different settings and scenarios. It is essential to communicate with the person and pay attention to their feedback. Moreover, when we are developing new solutions, activities, and spaces, we should invite them to participate and contribute. They have a lot to offer from their unique point of view, expertise, and insight. He shared,

> "It is about empathetically considering the experiences of those with disabilities," Gary points out. "You've got to have those conversations with the person and actually listen to what they are saying."

When those impacted the most are actively involved and given a voice in the design process, their unique insights and expertise can make the design more appealing to a broader audience. This inclusive participative approach, coupled with Universal Design Principles, can lead to exceptional outcomes.

On a positive note, as Gary put it, the trend in our society is that we are becoming more aware and inclusive of people with disabilities. They are no longer hidden or isolated, but rather participate actively and confidently in various aspects of life. We can notice them more frequently in the media, not only as inspiring role models who defy the challenges they face but also as ordinary people who have diverse personalities and interests. Their disability is not the main focus of their story but simply one of the many facets of their identity. This helps to reduce the fear or awkwardness that some people may experience when they encounter someone with a disability and encourages them to treat them with respect and empathy.

EDUCATE

Beyond inclusion within the design phase, education is the next best defense. Gary suggested one way to raise awareness is to incorporate it into the educational framework of the various engineering fields. He adds,

> Imagine the impact if we could also introduce a course focused on the viewpoints of individuals with disabilities! Occasionally, we witness this in action during senior design team projects, where a handful of students come up with a creative solution for a person, frequently a child, who has a disability. The positive impact of such projects can profoundly affect both the recipient and the students involved. It is my hope that more engineers have the opportunity to engage in such rewarding experiences.

Gary was amazed at how technology can evolve and transform over time. He recalls that not so long ago, there were "new" innovations that aimed to assist people with disabilities in their writing and communication tasks. These innovations were speech to text software and word prediction, which enabled users to convert their voice into text and get suggestions for the next word based on the context. These tools were very helpful and empowering for people who faced challenges with typing or expressing themselves. However, today these tools are no longer exclusive to the disability community. They have become essential features of every smartphone that exists. Many people use them on a daily basis, either for convenience, speed, or accuracy. They may not even realize that these tools originated from the disability sector, and that they are benefiting from the efforts and needs of that group. This shows how technology can have a wide and unexpected impact on society, and how it can cross boundaries and reach new audiences.

Technology has played a key role in transforming the ergonomics of workplaces, especially for people with disabilities and older workers. These two groups have some common challenges, such as reduced strength, vision, hearing, and mobility, which may affect their performance and well-being at work. These challenges were often overlooked or ignored when the work environment or task was originally designed many years ago. However, technology has provided solutions that enable these workers to overcome their limitations and work more comfortably and efficiently. Some of these solutions include tool balancers, lighter weight power tools, multi-fingered power tools, push-assisted carts, font size adjustments on screens and labels, multimode signals or warnings that combine visual and auditory cues, and cobots that assist workers with their tasks. Gary noted that as these technologies become more advanced and accessible, they will be more widely adopted in various workplaces, especially cyber-physical workspaces.

CONSIDERATIONS FOR PERSONS WITH COGNITIVE DISABILITIES

While ergonomics practitioners are often primarily tasked with designing for individuals with physical disabilities, having expertise in providing guidance for those with cognitive disabilities is also highly beneficial. Neurodivergent individuals (explored further in the next chapter), who make up approximately 15-20% of the global population, include conditions such as Autism Spectrum Disorder and Attention-Deficit/Hyperactivity Disorder. These conditions, considered as an "invisible disability" as their symptoms are not immediately apparent or visibly obvious to others (Invisible Disabilities Association, n.d.), can overlap with mental disabilities, particularly in terms of their effects on daily functioning and workplace performance.

Accommodating both physical and cognitive disabilities is not just about fostering inclusion; it also offers significant economic and organizational advantages. Businesses and organizations that invest in creating accessible environments for individuals with diverse needs can benefit from various global incentives and support programs. For instance, many countries provide tax credits, grants, and funding opportunities to encourage businesses to invest in accessibility and accommodations. Here are some examples:

- United States: Work Opportunity Tax Credit; Disabled Access Credit.
- Barrier Removal Tax Deduction.
- European Union: EU Disability Strategy.
- Australia: Employment Assistance Fund; Wage Subsidy Scheme.
- Canada: Opportunities Fund for Persons with Disabilities; Canada Disability Savings Program.
- United Kingdom: Access to Work Scheme.

In addition to these financial incentives, there are broader organizational benefits to consider. Creating accessible work environments allows businesses to tap into a larger talent pool, bringing in unique skills and perspectives that can enhance innovation and creativity. Designing for a diverse range of needs also promotes workplace flexibility and adaptability, making it easier to address future changes and challenges.

A blog article from Ripplematch (2020), a recruitment automation platform, shared examples of companies that have answered the call (specific to cognitive disabilities) such as L'Oréal U.S.A.'s employee-led resource group "DiversABILITY Think Tank," which focuses on developing accessibility initiatives. And then there are Walgreens' exemplary inclusive policies and practices in which they partner with community organizations, vocational rehabilitation agencies, and state and local social services agencies to identify work candidates for their Transitional Work Group Program and Retail Employees with Disabilities Initiative programs (Walgreens, n.d.). It is with these programs, cognitively disabled individuals are employed throughout the company's distribution network, retail stores, the support center, and field operations. With targeted career skills training, these individuals perform at the same high standards as other associates. This is because Walgreens believes that "*individuals with disabilities can do the same jobs for the same pay, and perform at the same high standards, as any other team members*," and they work to develop work processes and practices to provide opportunities so that they are able to do so.

As more organizations venture down this path of making work more accessible—possibly yours—this would be a great opportunity for ergonomics practitioners to share their expertise in helping to design work, workspaces, and work experiences to be inclusive and efficient for disabled individuals.

RETURN—TO WORK PROGRAMS AND DEI—KEVIN'S STORY

The week of January 8, 2018, started out as a normal, yet busy week for Kevin Doyle. As a seasoned business developer, he was immersed in his usual busy schedule of planning for innovation events, including attending the Design Thinking Gym in New York City and the Detroit Auto Show. To him, it was starting out to be a typical week of exciting work.

But it wasn't a normal week for Kevin. It was a week that changed his life as he knew it forever.

After a routine conference call, Kevin suddenly found himself inexplicably unable to use his keyboard or mouse. The actions became foreign to him. After walking into a cubicle wall soon after, he realized something was terribly wrong. And he was right—Kevin was having a stroke.

A subsequent hospital visit confirmed the cause: an Arterial Vein Malformation, a condition that had been undetected for years despite his ongoing struggles with headaches and seizures. Kevin underwent a craniotomy to address the malformation, but he knew that his life and work would never be the same. His neurosurgeon, Dr Sharon Webb, warned him that many stroke survivors never return to a fully productive work life. He shared,

> She pushed me, told it like it was and made sure from the beginning to expect challenges along the way. On my last call, she said that many people who experience this type of trauma never return to a fully productive work life.

Despite these challenges, Kevin was determined to return to the work he loved. He knew the journey would be difficult, but he didn't know just how much when it came to getting back to work. He faced cognitive and physical obstacles, including loss of peripheral vision and balance issues, and struggled with a lack of confidence in front of others. Kevin admits,

> I knew I was not as comfortable in front of people but I did not want to be treated differently. Instead suffered on my own, I should have leaned on the organization for help to ease back into the role and to see what we might put in place to get me feeling more like myself.

Kevin's experience highlights the broader challenges for many people returning to work after an acquired brain injury, such as a stroke. Strokes are recognized as the single largest cause of severe disability globally, with 60% of stroke survivors needing job modifications after a stroke (Tsutsumi, 2020). Medin et al. (2006) identified that many stroke survivors perceived rehabilitation as focusing mainly on bodily functions and carrying out everyday activities rather than promoting a return to work.

With the rates of stroke among younger adults on the rise (according to a new report from the U.S.'s CDC), this is a pertinent issue for the workplace today (CDC, 2024b). Pearce et al. (2023) noted the growing concern about increased stroke events among younger adults, who may face significant barriers when attempting to return to the workforce after a stroke and who may not have options, such as early retirement, that older workers have. They also identified that interventions aimed at promoting return to work after stroke have been limited.

Return-to-Work (RTW) programs are structured initiatives designed to assist employees in recovering from an injury or disability and successfully reintegrating into the workplace. These programs are based on a medical assessment conducted by a healthcare professional, leading to the creation of a customized plan that outlines a timeline and specific recovery steps. The plan may include transitional duties or lighter tasks that gradually increase in complexity as the employee's strength and capabilities improve.

In addition to these recovery steps, RTW programs provide supportive measures, such as workplace accommodations (e.g., workspace adjustments, flexible work hours, or assistive technology) and training for both the employee and their colleagues to ensure a smooth transition back to work. These programs are crucial for fostering inclusion and equity within organizations by offering tailored support to employees recovering from health challenges, ensuring they have equal opportunities to re-enter the workforce. Return-to-Work programs also help organizations retain valuable employees who might otherwise leave due to their health issues, thus maintaining a diverse and talented workforce.

The ergonomics practitioner's role in a comprehensive RTW program involves several key actions to ensure employees can return to their roles safely and effectively while managing their condition. This includes conducting workplace assessments to identify potential hazards and areas that need modification, as well as analyzing job tasks to determine necessary adjustments that align with the employee's current abilities. We will recommend changes to the workstation, specialized tools, or equipment to improve comfort and productivity, as well as help to identify modified job responsibilities, restricted duties, or temporary job transfers to accommodate the employees" needs following an injury or disability. A key part of the process also involves collaborating with healthcare providers to ensure they understand the workplace tasks and conditions, which allows for the creation of tailored RTW plans that include a phased return to work, addressing the specific needs of the recovering employee. (CDC NIOSH, 2024a).

Returning to work after a physical injury, such as a broken leg or shoulder surgery, is often easier to manage because the recovery process can be seen as more predictable. The physical restrictions can be more clearly assessed and monitored, making it easier to plan appropriate RTW interventions, and the progress in physical recovery is typically visible and measurable, allowing for precise workplace or work task adjustments. However, when it comes to neurological conditions, such as those following a stroke, the challenges can be more complex and less immediately apparent. Cognitive disabilities and impairments, including difficulties with attention, memory, and executive function, are often "invisible" compared to physical injuries, making it more difficult for employers and colleagues to identify and accommodate the employee's needs. These needs may require highly individualized and flexible adjustments that could evolve over time. Therefore, a collaborative approach involving the employee and the accommodation team is crucial. This is also an opportunity for us as ergonomics practitioners to consider strategies for supporting neurodiverse individuals, as many of the cognitive challenges following a stroke, such as impacts on executive function, information processing, emotional regulation, and task performance, can overlap with those seen in neurodiverse conditions (discussed further in Chapter 9). Because of these similarities, the accommodations for both groups may share common elements that can be beneficial. The Stroke Foundation (n.d.) recommends the following interventions:

- **Flexible Work Hours ("Flex Work")**: Allow for working hours to be adjusted to match the employee's energy levels. This could include a later start, more frequent breaks, or working from home.
- **Assistive Technology**: Utilize assistive devices or software to aid with physical or cognitive challenges. This could include speech-to-text programs or screen readers.
- **Physical Workspace Adjustments**: Modify the workspace to accommodate any physical limitations. This might involve changing the layout, using adaptive equipment, or ensuring easy access to necessary resources.
- **Task Modification**: Break down tasks into smaller, more manageable steps. Prioritize tasks and focus on those that align with your strengths.

The American Stroke Association (n.d.) provides additional resources for organizations in preparing RTW strategies for employees returning to work after their stroke recovery. These resources include a decision tree that provides initial guidance and possible outcomes based on the desires of the employee, an employment readiness checklist that provides guidance on the employee's readiness based on ongoing symptoms and actions, and a goal-setting worksheet for the employee and organization to set objectives with metrics and target timelines.

In Kevin's experience, he believes he should have sought more support from his organization to help ease his transition back into his role and to identify accommodations that would help him feel more like himself again—a challenge that mirrors a factor that can make the accommodations process (a sister program) ineffective.

When asked what would have made his return to work more effective, Kevin pointed to a combination of physical and non-physical accommodations. Regaining confidence in social interactions was a major challenge for him after the stroke, especially given the customer-facing nature of his job. He felt that targeted training to boost his comfort and confidence in these interactions would have been particularly helpful, as it was an "essential" requirement of his role. Additionally, for his vision and balance difficulties, he wished he could have identified and implemented workspace modifications more quickly to address those issues.

Kevin stressed that returning to work after a brain injury, like a stroke, requires a personalized approach, as each individual's situation is unique and demands tailored support to effectively address their specific challenges. Looking back, he strongly encourages employees to familiarize themselves with their organization's RTW programs, as well as the accommodation process, understanding both the processes and the expectations of each. He believes that had he been more aware of these resources, it could have eased his own transition. He also feels that raising awareness of these programs can help other employees who are recovering from a stroke—or any brain injury—avoid facing their challenges alone and enable them to continue giving their best at work.

To this day, Kevin says that he hasn't regained 100% of his pre-stroke work confidence, but he recognizes how far he has come, even amazing his neurologist Dr Webb.

> When I get a win, it is a great day.

PRACTICALLY SPEAKING

Engaging with established processes and programs, such as the accommodations process and RTW initiatives, is essential for creating an inclusive workplace for employees with physical and cognitive disabilities. These programs offer the critical support needed for employees to perform their roles effectively. In addition to actively participating in these programs, Gary offers timeless advice for anyone involved in ergonomics that is both practical and profound:

- **Broaden Your Design Scope**: Don't limit yourself to just the 5th percentile female and the 95th percentile male. Instead, aim to accommodate a wider range of body sizes, shapes, and capabilities. This approach will help you create more inclusive and accessible solutions that serve a broader spectrum of individuals.
- **Learn More About Universal Design and Its Seven Principles**: Take the time to study Universal Design and familiarize yourself with its seven core principles. Universal Design focuses on creating environments and products that are accessible to everyone, regardless of their age, ability, or status. These principles include concepts like equitable use, flexibility in use, and perceptible information. By incorporating these principles into your work, you can design solutions that cater to diverse needs and ensure usability for all.
- **Engage with the Community**: Get to know people in your community, place of worship, or company who have disabilities. Talk to them and *actually* listen. Engaging in genuine conversations with them and actively listening to their experiences can provide invaluable insights. Understanding their real-world challenges and needs will help you design more effective and empathetic solutions.
- **Diversify Your Focus Groups**: When conducting focus groups, make sure to include people with various types of disabilities and capabilities. This diversity ensures that you gather a broad range of feedback and perspectives, which is crucial for designing inclusive solutions. Different disabilities—whether mobility, sensory, or cognitive—require different considerations, and including diverse participants will help you address these needs comprehensively.

I would agree that what Gary shared is great advice for any new ergonomics practitioner—or any seasoned one, for that matter. They encourage a mindset that prioritizes inclusion, empathy, and a deeper understanding of the diverse needs of all individuals, which is essential for effective ergonomics practice at any stage of one's career. Adopting this perspective helps ergonomics practitioners stay adaptable, relevant, and mindful of the evolving needs of the workplace and the people within it. And that remains pertinent at any stage of one's career.

FURTHER READING

For additional reading on designing and mitigating the challenges related to designing for persons with physical and cognitive disabilities, consider the following:

- *DEI Through Ergonomics: Supporting Diverse Body Types and Abilities* (Downey et al., 2023)
- The Centers for Disease Control and Prevention (CDC) Disability Inclusion blog (CDC, 2024c)
- United Nations Disability Inclusion Strategy (n.d.)

- Additional information on Web Content Accessibility Guidelines (WCAG) (n.d.)
- *Accessible America: A History of Disability and Design* by Bess Williamson (2019)
- Doing Disability Differently: An alternative handbook on architecture, dis/ability and designing for everyday life by Jos Boys (2014)

REFERENCES

American Stroke Association. (n.d.). Return to work. https://www.stroke.org/en/life-after-stroke/recovery/return-to-work

Boys, J. (2014). *Doing disability differently: An alternative handbook on architecture, dis/ability and designing for everyday life*. Routledge.

Centers for Disease Control and Prevention. (2024a). *Promote recovery*. National Institute for Occupational Safety and Health. https://www.cdc.gov/niosh/ergonomics/ergo-programs/promote-recovery.html

Centers for Disease Control and Prevention. (2024b). *COVID-19 vaccine administration and myocarditis: MMWR volume 73, report 20*. Centers for Disease Control and Prevention. https://www.cdc.gov/mmwr/volumes/73/wr/mm7320a1.htm

Centers for Disease Control and Prevention. (2024c). *About disability inclusion*. Centers for Disease Control and Prevention. https://www.cdc.gov/disability-inclusion/about/?CDC_AAref_Val=https://www.cdc.gov/ncbddd/disabilityandhealth/disability-inclusion.html

Downey, G., Gallo, E., & Stanch, P. (2023, June). *DEI through ergonomics*. American Industrial Hygiene Association. https://synergist.aiha.org/20230607-dei-through-ergonomics

Invisible Disabilities Association. (n.d.). *Defining invisible disability*. Invisible Disabilities Association. https://invisibledisabilities.org/ida-getting-the-word-out-about-invisible-disabilities/defining-invisible-disability/

Medin, J., Barajas, J., & Ekberg, K. (2006). Stroke patients' experiences of return to work. *Disability and Rehabilitation*, *28*(17), 1051–1060. https://doi.org/10.1080/09638280500494819

Pearce, G., O'Donnell, J., Pimentel, R., Blake, E., & Mackenzie, L. (2023). Interventions to facilitate return to work after stroke: A systematic review. *International Journal of Environmental Research and Public Health*, *20*(15), 6469. https://doi.org/10.3390/ijerph20156469

RippleMatch. (2020, October 21). *Companies with innovative and award-winning accessibility practices*. RippleMatch. https://ripplematch.com/insights/companies-with-innovative-and-award-winning-accessibility-practices-972ec8a4

Siler, B. (2018, April 12). 5 reasonable accommodation pitfalls to avoid. *HR Daily Advisor*. https://hrdailyadvisor.blr.com/2017/10/04/5-reasonable-accommodation-pitfalls-to-avoid/

Stroke Foundation. (n.d.). Reintegrating into the workforce after a stroke: A guide for survivors and caregivers. https://thestrokefoundation.org/reintegrating-into-the-workforce-after-a-stroke-a-guide-for-survivors-and-caregivers/

Tsutsumi, A. (2020). Return to work after stroke. In U. Bültmann & J. Siegrist (Eds.), *Handbook of disability, work and health* (Handbook Series in Occupational Health Sciences, Vol. 1). Springer. https://doi.org/10.1007/978-3-319-75381-2_25-1

United Nations. (n.d.). *United Nations disability inclusion Strategy.* United Nations. https://unsdg.un.org/resources/united-nations-disability-inclusion-strategy

Walgreens. (n.d.). *Disability inclusion at Walgreens.* Walgreens. Retrieved September 7, 2024, from https://www.walgreens.com/topic/sr/sr_disability-inclusion.jsp

Web Content Accessibility Guidelines (WCAG). (n.d.). Web Content Accessibility Guidelines (WCAG). https://www.wcag.com/

Williamson, B. (2019). *Accessible America: A history of disability and design.* NYU Press.

9 Designing for Neurodiversity

Picture it: You're sitting at your desk hammering away at your latest spreadsheet when you get a message from your immediate manager that simply states, "Call me." If you're anything like me, your mind immediately starts to race through all kinds of scenarios.

Did I forget an important task?
Did I flub my expense report?
Did I copy the wrong person in an email?
Did I forget to send an important email?
Will I have to look for another job?

On and on. When you finally calm down enough to actually give your manager a call (but not so calm that your heart isn't pounding so hard that you're sure it can be heard through the headset), they share some benign news about an upcoming event. They are just interested in knowing if you have the bandwidth to attend. That's it. So, it *wasn't* impending doom.

For those like me, anxiety is an emotion that can easily derail a day. And it is something that we must be intentional about managing. But it is also something that others we work with and for should be mindful of as well—because our actions can significantly impact people in ways we may not be aware of. This is because many people are neurodivergent.

"Neurodiversity" typically refers to the natural variation in how our brains cognitively function, including different ways of thinking, learning, and processing information. "Neurodivergent" refers to cognitive functions that vary from what is considered "typical" (Goldberg, 2023)—such as autism spectrum disorder, attention-deficit/hyperactivity disorder (ADHD), dyslexia, and other neurological or developmental conditions. Being neurodiverse is not considered having a disability, just a different way of functioning—another characteristic of diversity that should be understood and accounted for to help ensure all are able to contribute in their best way.

If we look back at the "call me" situation, it is an example of how the work environment, including how we interact with others, can become an imbalanced system and lead to psychosocial stress. I spoke with Victoria Ballard, a Certified Professional Ergonomist and Professional Engineer with a PhD in Industrial and Systems Engineering from Auburn University concentrating on Occupational Safety and Ergonomics and Manufacturing Systems, to discuss the importance of why understanding neurodiversity is critical for our ergonomics profession. She explained

DOI: 10.1201/9781032658988-12

that from a design aspect, even this simple act of how we communicate is an example of what can happen when neurodiversity isn't considered. She says,

> For some individuals, these vague prompts can trigger significant anxiety, leading to a flood of worrisome thoughts such as "What did I do wrong?," "Did I forget something important?," or even "Am I about to be fired?" This anxiety can become all-consuming, especially if the recipient is working in environments that require high concentration.

Hello! That was me!

Victoria shared how minor adjustments to how we frame our communication can be effective. Instead of sending a brief "call me" message, providing some context in the message can substantially alleviate anxiety. For example, a message saying, "When you have a minute, please call me to discuss the logistics for tomorrow's trip, not urgent. Thanks," gives the recipient a clear idea of the meeting's subject and its urgency level. She shared this practice as an example of designing for neurodiversity. It not only fosters a more inclusive and supportive work environment but also ensures that all employees, regardless of their neurodiverse status, can maintain focus and efficiency without unnecessary stress.

But why is this important? Breslin and Pole (2009) conducted research that speaks to how these diverse cognitive styles present challenges for individuals in the workplace, which could potentially elevate the risk of accidents and negatively impact quality and productivity. When employers endeavor to enhance workplace safety and efficiency, their decisions are primarily shaped by the needs and characteristics of neurotypical employees—with neurotypical referring to individuals who demonstrate a normative pattern of neurodevelopment (Chown, 2021). And the simple communication example that Victoria shared illustrates how we as ergonomics practitioners, can easily support and promote psychological safety in the workplace.

PSYCHOLOGICAL SAFETY AND NEURODIVERSITY

In talking about the history and focus on neurodiversity, Victoria shared how the term was originally coined by sociologist Judy Singer as part of rights activism aimed at recognizing and respecting naturally occurring cognitive differences among individuals (Singer, 1999). The neurodiversity movement initially highlighted the positive aspects of these differences, such as creativity and specialized skills, rather than focusing solely on deficits or disabilities associated with neurocognitive developmental disorders. However, over time, the understanding of neurodiversity has broadened from focusing on specific conditions to embracing a wider spectrum that includes all forms of cognitive diversity. The term has evolved from being associated with disability to a more inclusive concept that encompasses any individual's cognitive differences, with terms like "neurodivergent" and "neuro-minority" being used to describe individuals who differ from neurotypical norms (Doyle & McDowall, 2021).

Research in this space, including Victoria's, has shown that despite the progressive intentions behind the neurodiversity movement, focusing on neurodiversity in professional and social contexts poses several challenges. For example, Victoria has observed a significant lack of research and practical guidance concerning neurodiversity in the workplace that could help employers and employees manage and benefit from it (Ballard, 2022, 2024). This gap leads to challenges in developing effective policies and practices that support neurodivergent individuals.

There is also an identified gap with respect to applied research in industry. Victoria called this the "Science-Practitioner Gap," where there is a noted disconnect between scientific research and its practical application, particularly in organizational contexts. While there is growing recognition of the potential contributions of neurodiverse individuals to the workforce, there is insufficient evidence-based guidance on how to integrate this potential into existing human resource practices effectively. Hence another reason for this book! This is critical considering that studies have reported neurodivergent workers as having significantly higher rates of unemployment, sick days, workplace injuries, and workplace conflicts. Researchers call for consideration of psychosocial programs tailored to these workers' needs (Das et al., 2012; Gémes et al., 2022; Khan et al., 2022).

Beyond the lack of practical guidance for industry, Victoria shared that the prevailing medical model often pathologized neurodiverse conditions, focusing on deficits rather than strengths. This perspective can hinder the full integration and acceptance of neurodivergent individuals in the workplace, as it promotes a view of these individuals as needing "treatment" or "correction" rather than recognizing and utilizing their unique skills. It is the unfortunate "othering" we see throughout discussions around diversity, equity, and inclusion that can unintentionally sideline Diversity, Equity, and Inclusion (DEI) efforts (Mackenzie & Abad, 2021). This, unfortunately, can also lead to neurodivergent individuals facing significant social and legal barriers, such as discriminatory hiring practices and workplace accommodations that are not suited to their unique needs, a point mentioned in the previous chapter when discussing challenges with accommodations processes. These barriers can lead to underemployment and unemployment among neurodiverse populations, perpetuating social and economic exclusion (Thomas & Morgan, 2021).

And lastly, workplace norms and culture are often designed around the needs and behaviors of neurotypical individuals. This can create environments that are inherently unwelcoming or inaccessible to neurodivergent individuals, who may have different ways of processing information and interacting socially (Sundar, 2017).

Victoria noted that, in her perspective, it will take a concerted effort to address these challenges and bridge the gap between research and practice, develop inclusive policies that recognize neurodiversity as a form of diversity, and create workplace cultures that genuinely value and accommodate all forms of cognitive difference.

So, where do we begin?

We begin by understanding the importance of psychological safety with respect to neurodiversity. Victoria has observed, through attending conferences such as the Applied Ergonomics Conference and those held by the Human Factors and Ergonomics Society and American Society of Safety Professionals, that there is a notable shift in discussions toward psychological safety and addressing the needs of neurodiverse populations.

Referring back to what was discussed earlier about psychological safety, it is about allowing employees to feel comfortable expressing their opinions, asking for help, admitting errors, and challenging the status quo without fear of negative consequences. It can foster trust, respect, and inclusion. For employees who are neurodiverse, psychological safety is critical because their diverse ways of thinking and responding to situations can mean that they experience higher levels of stress and frustration in the workplace and can face various challenges such as misunderstanding, isolation, or exclusion. All of these factors can significantly impact their well-being, productivity, and engagement at work.

Victoria has also noticed how research has begun to draw strong connections between psychological safety and musculoskeletal disorders and injuries, supporting a trend toward creating psychologically safer workplaces for all workers, including those who are neurodiverse. However, while she has recognized that this trend is emerging in our profession and adjacent professions, it is less apparent in manufacturing and other sectors, and that focus is needed there. She added,

> It is an exhilarating time to be involved in this field, as there is much to learn about how human factors and ergonomic principles intersect with cognitive processing and physical responses. Exploring how to best accommodate workers who process and sense the world differently is still a relatively new area of discussion. Not all circles are engaged yet, but I anticipate it will become more widespread as our understanding deepens.

Victoria went on to add that this evolution in the field of ergonomics marks a significant shift. As Gary shared with us in design considerations for physically disabled individuals, Victoria noted that the same is true for those who are neurodivergent—with a shift from designing systems solely for "normal" or "neurotypical" users, as traditionally outlined in our textbooks, to adopting a Universal Design approach. This approach aims to create systems that are safe and usable for a broader spectrum of users.

CHALLENGES WHEN DESIGNING FOR NEURODIVERSITY

Similarly to the challenges of implementing accommodations for persons with physical and cognitive disabilities, Victoria noted that there are also challenges with respect to accommodations for those who are neurodiverse. The main challenge is often the workplace culture. If a workplace's culture doesn't provide a safe and open environment for discussing neurodivergent traits or disabilities, then recognizing needs and implementing those needs can be met with resistance.

Another challenge is perception. Without understanding the conditions a person may have or awareness of how to recognize how challenges exhibit themselves, it is possible for others to wrongly label traits of a condition as something else—such as laziness or a person just being difficult to work with. For example, those with ADHD can exhibit behaviors that are socially challenging and may be misinterpreted by others. This is due to a dysfunction of the "executive functions"—a group of processes that include working memory, cognitive flexibility, and inhibition control, as defined by the Cleveland Clinic (Cleveland Clinic, n.d.). This could mean that people can perceive those with ADHD as "unmotivated," "uncaring," or "sloppy" (to name a few) when in actuality it could be due to the person being overstimulated, anxious, or overwhelmed.

Another challenge is when people who are neurodivergent "mask" their conditions. Masking is when a neurodivergent individual conceals their behavior to conform to neurotypical standards. They may mask to gain social acceptance or simply to try and navigate social situations. However, constant masking can lead to emotional strain, stress, and anxiety—all potentially increasing the risk potential for unsafe acts and (Beck, 2023) musculoskeletal pain (Ryan et al., 2023). Not feeling comfortable enough to share their true authentic self is also an indication of a lack of psychological safety within that individual's workplace (or other social environments).

Promoting Neurodiversity Inclusion

In terms of actionable strategies to promote neurodivergent inclusion in workplaces, Victoria stresses a caveat: While some neurodivergent conditions share traits, many have distinctive challenges and strengths. In some cases a modification to help one neurodivergent individual may cause negative impacts for another. A best practice is to have a workplace culture that supports the communication of needs of workers to identify what works best (remember what I said about Participatory Ergonomics?) See Table 9.1 for more suggested strategies as shared by Victoria.

In terms of advice Victoria would like to share for ergonomics practitioners, as well as companies deploying DEI strategies to enhance the inclusiveness of employees who may be neurodivergent, it is important to first become aware of what it means to be neurodivergent and then promote that understanding with others. This means understanding neurodiversity encompasses a range of conditions—including autism, ADHD, dyslexia, and others—and that each condition affects individuals differently. Therefore, workplace accommodations must be flexible and personalized, not one-size-fits-all.

She also recommends, in terms of designing for the workplace, consideration should be given to workspace designs that support various sensory sensitivities. For example, options for adjustable lighting, noise control with noise-canceling headphones, and flexible seating options such as standing desks or areas where employees can retreat for quiet time if needed. In terms of digital workspaces, providing software that supports different learning and processing styles, such as text-to-speech for those with dyslexia or time management tools for those with ADHD, can be immensely beneficial. And in terms of communications and training materials, they should be accessible—utilizing clear language, subtitles, and visual aids to cater to diverse cognitive needs. All these solutions that could prove beneficial for neurotypical workers as well.

And speaking of training, Victoria highly recommended that any DEI initiative include training for all employees on the basics of neurodiversity. She strongly believes that understanding what neurodiversity is and recognizing the strengths and challenges of neurodivergent individuals will help to foster a more supportive and collaborative work environment as well as help to reduce stigma. And most

TABLE 9.1
Suggested Strategies for Neurodiverse Employees as Shared by Victoria Ballard

Strategy	Examples
Education and Awareness Training	• Conduct diversity and inclusion training for all employees to raise awareness of different abilities, including neurodiversity. • Provide training on unconscious bias to help employees recognize and overcome preconceived notions about colleagues.
Accommodations and Accessibility	• Ensure that the physical workplace is accessible to all employees, including those with mobility challenges. • Implement accessible technology and tools that accommodate various abilities. • Develop clear procedures for requesting and providing reasonable accommodations.
Inclusive Recruitment and Hiring	• Revise job descriptions and qualifications to focus on essential skills and competencies, reducing potential biases that could exclude qualified candidates. • Actively recruit from diverse talent pools, including organizations that support neurodivergent individuals. • Partner with community organizations and vocational programs to identify and hire neurodivergent talent.
Mentorship and Support Programs	• Establish mentorship programs where experienced employees can guide and support neurodivergent colleagues. • Create employee resource groups or affinity groups that provide a sense of community and support for neurodivergent individuals.
Communication Accessibility	• Ensure that all written and digital communication materials are accessible, including formats suitable for individuals with visual or cognitive impairments. • Encourage clear and straightforward communication in the workplace to benefit all employees.
Flexible Work Arrangements	• Offer flexible work schedules or remote work options where feasible to accommodate diverse needs. • Allow employees to adjust their workstations or environments to suit their preferences and requirements.
Sensory-Friendly Workspaces	• Design workspaces with attention to sensory needs, such as reducing noise levels or providing quiet spaces for breaks. • Consider the lighting, temperature, and layout to create a comfortable environment.
Feedback and Continuous Improvement	• Solicit feedback from neurodivergent employees and use it to make necessary improvements in policies and practices. • Regularly assess the effectiveness of inclusion initiatives and adjust them as needed.
Inclusive Leadership and Role Models	• Promote diversity in leadership positions to serve as role models for all employees. • Encourage leaders to demonstrate inclusive behaviors and prioritize diversity and inclusion in decision-making.
Conflict Resolution and Mediation	• Establish a fair and transparent process for addressing workplace conflicts and grievances. • Provide mediation services to help employees resolve disputes amicably.

(*Continued*)

TABLE 9.1 (CONTINUED)
Suggested Strategies for Neurodiverse Employees as Shared by Victoria Ballard

Strategy	Examples
Inclusive Language and Behavior	• Promote the use of inclusive language and discourage derogatory or exclusionary language and behavior. • Encourage employees to speak up if they witness discriminatory or disrespectful actions.
Celebrate Diversity and Achievements	• Recognize and celebrate the achievements and contributions of all employees, regardless of their background or abilities. • Organize events and initiatives that highlight diversity and foster a sense of belonging.
Regular Check-Ins and Feedback	• Conduct regular one-on-one meetings with employees to discuss their needs, challenges, and career aspirations. • Provide a feedback mechanism for employees to voice concerns and suggestions anonymously if desired.

importantly, it will help to promote psychological safety for not just neurodivergent employees, but all employees. And by tapping into the participatory approach to ergonomics and DEI, particularly through Employee Resource Groups, companies can establish integrated feedback mechanisms. These mechanisms can create a supportive environment for employees facing challenges, identify potential workplace issues, and offer practical evaluations of proposed solutions.

Neurodiversity and the Accommodations Process

Returning to my conversation with Ann Hall, she emphasized the critical role of ergonomics practitioners in the accommodation process, particularly for neurodiverse employees. She highlighted that such accommodations not only support the individual but also benefit employers by improving overall productivity and retention in the workplace. The process of making adjustments for employees with disabilities or specific needs, she noted, is a win-win scenario when approached thoughtfully and inclusively.

Ann shared an example of her work with an employee who was on the Autism Spectrum and had ADHD. Research from the National Library of Medicine shows that elevated background noise can significantly worsen symptoms of inattention and ADHD, making it a key consideration in workplace accommodations. Furthermore, the Northwestern University Early Intervention Research Group has identified hyperacusis, or heightened sensitivity to sound, as a common issue for individuals with autism. In this particular case, the employee struggled in a bullpen-style office environment in the middle of a manufacturing floor, where excessive noise from drilling, riveting, and jackhammering, proved highly distracting. While Ann's primary goal was to find a solution that would accommodate the employee with ADHD, she also recognized the broader need to improve the work environment for the entire team, since the noise was affecting everyone, not just the individual. Ann shared,

> In my experience, issues that neurodiverse employees are encountering at work are also impacting other employees, just not with the same magnitude. At one site, employees were standing on desks to tape paper over the new fluorescent lighting. A couple of people were more bothered with severe headaches, but all complained about being bothered and constantly distracted. The situation with the noise was much the same. Our neurodiverse employee was experiencing PTSD type syndromes, but all were generally distracted and disturbed by the constant noise. Any improvements would have impacted the employee experience for everyone. Many had to reserve a conference room or find another desk just to have a simple phone conversation or online meeting.

To address the issue, Ann researched practical solutions for individuals with sound sensitivities, many of which were outlined by autism.org. These strategies included wearing earplugs or noise-canceling headphones in noisy environments and setting up workspaces with higher or closed walls to reduce noise exposure.

> We looked at sound blocking curtains and insulation, several different solutions that would benefit the whole department. The goal was to not design for one person, but the entire team. As a quick fix while we were figuring out a more long-term plan, the employee was fitted with custom earplugs, which he really liked because they were discreet and helpful in blocking the noise. Two weeks later, I was approached by the manager with a request for these ear plugs for the entire team.

When researching the issue, Ann found it useful to consider and share other recommendations from autism.org, such as using light covers, sunglasses, or hats to mitigate fluorescent lighting in the physical work environment. Also, there were interesting recommendations to avoid strong scents, select foods that avoided discomforting textures, choose soft, comfortable clothing, and adjusting work schedules to avoid peak hours or crowded spaces.

Unfortunately, not all employees feel comfortable seeking accommodations, even when they would benefit from them. A 2024 report from Inclusively, State of Workplace Inclusion & Accommodations (2024), reveals that nearly 43% of the 9,000 job seekers and current employees surveyed felt unsafe asking their employers for disability accommodations. Of those, 55% cited mental health conditions as a primary reason for needing support, while 49% identified as having learning disabilities, ADHD, or other neurodiverse conditions. These statistics highlight the urgent need for workplaces to not only expand their knowledge and skill sets for providing accommodations but also foster a more inclusive and psychologically safe environment where all employees feel comfortable seeking the support they need.

PRACTICALLY SPEAKING

Just as my enthusiasm for the field was ignited by a transformative Human Factors course at Clemson University, Victoria's journey into this research area also began in a Human Factors class, this time at Auburn University (kudos to Human Factors courses!). While attending, she recognized that the standard guidelines for how people typically process and perceive information didn't reflect her experiences—or those of many others, she suspected. This realization motivated her PhD dissertation research, which explores how adjustments to a manufacturing worker's workstation can affect individuals, especially those who are neurodiverse. She particularly focused on workers with ADHD, given its prevalence and its effects on executive functioning, memory, and attention. Victoria reflects on her experience, stating,

> Being at the forefront of investigating human factors in relation to neurodiverse workers has been immensely rewarding and challenging. With limited existing research to build upon, our work has opened the door for future studies that can expand on our foundational findings.

The increasing visibility of similar research themes at conferences and in publications makes her optimistic about the ongoing exploration and importance of this vital area. And I, too, share Victoria's optimism and look forward to what more we can learn and then apply as ergonomics practitioners in this space. In the interim, we can always take lessons from practical solutions such as those shared by Ann and from culture-developing initiatives currently being implemented across organizations today who recognize the importance of making the workplace inclusive for those who are neurodivergent. An article from Forbes magazine (Raz,2024) highlighted the following companies and their efforts to be neuroinclusive:

- Ernst and Young established 23 Neurodiverse Centers of Excellence (NCoEs) globally, focusing on fields like analytics and artificial intelligence. These centers have led to increased productivity, increased retention rates for neurodivergent employees, and over $1 billion in value creation.
- Microsoft implemented a Neurodiversity Hiring Program, which features an AI-driven job-matching platform and a developed interview process tailored for neurodivergent candidates. Microsoft has achieved a 97% retention rate for candidates placed through this program, resulting in higher job satisfaction.
- JPMorgan Chase's Autism at Work program focuses on hiring individuals with autism, providing customized onboarding and ongoing support. Employees in this program have been reported to be 90–140% more productive and make fewer errors compared to their neurotypical counterparts.

Creating an inclusive work environment for neurodivergent individuals does not always require a dedicated team, program, or initiative, as shown by the examples of the companies mentioned earlier. Even small efforts can significantly enhance employees' sense of inclusion and foster a thriving workplace. As with the communication example, simply adding more context to a short message can go a long way to minimizing stress from some individuals (and let's face it, minimize confusion all around).

In Chapter 12, I provide further guidance on how to design communications for a diverse workforce, including tailored recommendations for those with neurodivergent conditions like dyslexia and dyscalculia. However, beyond these specific guidelines, the most effective approach, as discussed in PART I, is to adopt an inclusive mindset from the very beginning of the design process. This is particularly important when considering neurodiversity, given its vast and complex nature. And as with the concept of "designing for diversity," there can't be one definitive checklist. That would be impossible. But having a mindset of recognizing potential barriers for those who may be neurodivergent (such as challenges individuals with dyslexia may have with interacting with wayfinding signs or how office lighting and noise can hinder the concentration of certain individuals) and accounting for them in the design phase is the best approach.

Engaging the workforce for feedback is also essential. From personal experience, during the design of a new office complex, we consistently collaborated with an employee-driven business resource group (BRG) that included members who were neurodivergent, or were allies of cognitive and physically disabled individuals. This collaboration allowed us to gather valuable insights that shaped the final design, resulting in features such as dedicated concentration areas with minimal distractions, lactation rooms, and improved ambient lighting—all based on feedback from various BRGs. By prioritizing inclusion, being intentional in recognizing and understanding design barriers for those who are neurodivergent and actively engaging with neurodivergent voices in the design process, organizations can create environments where everyone can thrive and contribute their best.

FURTHER READING

For additional reading on designing and mitigating challenges related to neurodiversity, consider the following:

- Job Accommodation Network (JAN): The JAN offers guidance on workplace accommodations and disability employment issues, helping employers and individuals understand how to support neurodivergent workers effectively. (askjan.org)
- Mental Health at Work: This U.K.-based initiative provides resources specifically aimed at improving mental health in the workplace, which includes supporting neurodiverse individuals. (https://www.mentalhealthatwork.org.uk/)
- Understood: This website focuses on learning and attention issues, offering comprehensive articles, tools, and expert advice on managing and supporting neurodivergence both in educational settings and the workplace. (https://www.understood.org/)

REFERENCES

Ballard, V. (2022). *Impacts of workers with attention deficit hyperactivity disorder. Written for PSYC 7700, Dr.* Jesse Michel, Auburn University Course.

Ballard, V. (2024). *Assessing technology effects on neurotypical and ADHD manufacturing workers' mental workload and performance and impact of workplace support structures on sense of self-efficacy and social anxiety* [Doctoral thesis]. Auburn University.

Beck, K. B., MacKenzie, K. T., Kumar, T., Breitenfeldt, K. E., Chang, J. C., Conner, C. M., Mandell, D. L., White, S. W., & Mazefsky, C. A. (2023). "The world's really not set up for the neurodivergent person": Understanding emotion dysregulation from the perspective of autistic adults. *Autism in Adulthood*, *5*(1), 1–11. https://doi.org/10.1089/aut.2023.0214

Breslin, F. C., & Pole, J. D. (2009). Work injury risk among young people with learning disabilities and attention-deficit/hyperactivity disorder in Canada. *American Journal of Public Health*, *99*(8), 1423–1430. https://doi.org/10.2105/AJPH.2008.140855

Chown, N. (2021). Neurodiversity. In F. R. Volkmar (Ed.), Encyclopedia of Autism Spectrum Disorders (pp. 1–5). Springer. https://doi.org/10.1007/978-3-319-91280-6_102298

Cleveland Clinic. (n.d.). Executive function. *Cleveland Clinic*. https://my.clevelandclinic.org/health/articles/executive-function

Das, D., Cherbuin, N., Butterworth, P., Anstey, K. J., & Easteal, S. (2012). A population-based study of attention deficit/hyperactivity disorder symptoms and associated impairment in middle-aged adults. *PLoS One*, *7*(2), e31500. https://doi.org/10.1371/journal.pone.0031500

Doyle, N., & McDowall, A. (2021). Diamond in the rough? An "empty review" of research into "neurodiversity" and a road map for developing the inclusion agenda. *Equality, Diversity and Inclusion: An International Journal*, *41*(3), 352–382. https://doi.org/10.1108/EDI-06-2020-0172

Gémes, K., Björkenstam, E., Rahman, S., Gustafsson, K., Taipale, T., Tanskanen, A., Ekselius, L., Mittendorfer-Rutz, E., & Helgesson, M. (2022). Occupational branch and labor market marginalization among young employees with adult onset of attention deficit hyperactivity disorder—a population-based matched cohort study. *International Journal of Environmental Research and Public Health*, *19*, 7254–7254. https://doi.org/10.3390/ijerph19127254

Goldberg, H. (2023). Unraveling neurodiversity: Insights from neuroscientific perspectives. Encyclopedia, *3*(3), 972–980. https://doi.org/10.3390/encyclopedia3030070

Inclusively. (2024). The state of workplace inclusion and accommodations report. *Inclusively*. https://www.inclusively.com/wp-content/uploads/2024/11/Inclusively_-The-State-of-Workplace-Inclusion-Accommodations_a11y.pdf

Mackenzie, L., & Abad, M. (2021, February 5). Are your diversity efforts othering underrepresented groups? *Harvard Business Review*. https://hbr.org/2021/02/are-your-diversity-efforts-othering-underrepresented-groups

Khan, M. H., Grabarski, M. K., Ali, M., & Buckmaster, S. (2022). Insights into creating and managing an inclusive neurodiverse workplace for positive outcomes: A multistaged theoretical framework. *Group & Organization Management*, 10596011221133583. https://doi.org/10.1177/10596011221133583

Raz, M. (2024, October 11). How top companies are leading the charge for a neuroinclusive workplace. *Forbes*. https://www.forbes.com/councils/forbescoachescouncil/2024/10/11/how-top-companies-are-leading-the-charge-for-a-neuroinclusive-workplace

Ryan, L., Beer, H., Thomson, E., Philcox, E., & Kelly, C. (2023). Autistic traits correlate with chronic musculoskeletal pain: A self-selected population-based survey. *OBM Neurobiology*, *7*(1), 155. https://doi.org/10.21926/obm.neurobiol.2301155

Singer, J. (1999). "Why can't you be normal for once in your life?" From a problem with no name to the emergence of a new category of difference. In S. French & M. Corker (Eds.), *Disability discourse* (pp. 59–67). Open University Press. https://neurodiversity2.blogspot.com/p/blog-page.html

Sundar, V. (2017). Operationalizing workplace accommodations for individuals with disabilities: A scoping review. *Work: Journal of Prevention, Assessment & Rehabilitation*, *56*(1), 135–155. https://doi.org/10.3233/WOR-162472

Thomas, F., & Morgan, R. L. (2021). Evidence-based job retention interventions for people with disabilities: A narrative literature review. *Journal of Vocational Rehabilitation*, *54*(2), 89–101. https://doi.org/10.3233/JVR-201122

10 Designing for Varying Generations

And the crowd went wild!

At least, that is how I remember the "Generation Challenge" I hosted during my keynote speech at the 2022 Applied Ergonomics Conference. To kick things off, I invited the diverse audience of ergonomics practitioners to cheer on their respective generations in a lively and competitive spirit.

It was fascinating to see how quickly we can become passionate (and, dare I say, a bit cliquish) when discussing generational differences. However, I couldn't help but wonder if I had inadvertently fallen into the trap of relying on the stereotypes that often dominate social media narratives—like the infamous "millennials and their avocado toast" memes that seem to be everywhere. As I engaged with the audience, I reflected on the importance of moving beyond these clichés to foster a more nuanced understanding of the unique strengths and contributions each generation brings to the table.

For over three decades, Gary Allread conducted ergonomics research and provided ergonomics technical assistance to companies nationwide while at The Ohio State University (yes, he made me put "The"). Gary is a Certified Professional Ergonomist, whose interest in the field was sparked by his personal experiences with physical ergonomics and efficiency. As both a left-handed and tall (over the 95th percentile) individual, he became acutely aware of how poorly designed environments can affect people's lives. Motivated by this realization, he chose to pursue a PhD to deepen his knowledge of human factors and ergonomics, aiming to apply his insights to improve the world around him.

I spoke to Gary to get his perspective on designing for the various generations. To him, a focus on this means we're taking the time to understand that the respective experiences each person has, regardless of their generation, impact their views of the world, work, and society. He feels that as ergonomics practitioners, it is important for us to understand the rationale behind differing opinions and attitudes, to avoid making assumptions or having prejudices that fault the employee rather than work design as the reason for ergonomics-related issues. He added,

> …the opinions and attitudes of employees from different generations are not right or wrong, they are shaped by the experiences and events occurring during their era of development.

Those differing experiences and perspectives translate into differing voices that can be used to collaborate for process improvement and innovation. Additionally,

DOI: 10.1201/9781032658988-13

recognizing the knowledge each of the varying generations can bring can significantly improve the workplace. However, it is very easy for us to fall into the stereotype trap. These stereotypes can lead us to make incorrect assumptions and conclusions, leading to potential bias that can create barriers (for ourselves as well as others as it prevents us from truly learning and understanding the other person).

STEREOTYPES

Gary recognizes several ingrained assumptions, stereotypes, and misunderstandings about those in different generations. He added,

> Comments I often hear from clients: "This younger generation just doesn't want to work, pay their dues, work their way up, play on their phones, etc."

Gary shared that generational stereotypes often paint a broad picture of each group, but they don't always reflect the full story or are completely wrong in their assumptions about an individual. As far as other generalizations of general traits and tendencies for each generation, Gary added that Traditionalists (born 1928–1945) are often seen as resistant to change, preferring stability and sticking to "old-school" work methods. Baby Boomers (1946–1964) are typically viewed as competitive and focused on individual success, with a reluctance to embrace new technologies or modern workplace practices. Generation X (1965–1980) is sometimes labeled as cynical and disengaged, valuing life-work balance over company loyalty and preferring to work independently. Millennials (1981–1996), beyond being seen as having, as he noted before, poor work ethics, are often criticized for being entitled, self-absorbed, and a lack of loyalty to employers. Generation Z (1997–2012) is sometimes seen as easily distracted, overly reliant on technology, and lacking people skills, with a preference for digital communication over face-to-face interaction.

These stereotypes are generalizations and don't capture the full complexity of each generation's contributions to the workplace. And while it may be fun to have a "us vs them" mentality when it comes to repping our generations, we have to remember, as with any ergonomics consideration, that each person is unique and it is neither fair nor productive to attribute a blanket characteristic to a person simply because of the year they were born.

GENERATIONAL AWARENESS

However, we can recognize how the potential exposure to significant experiences and events, such as political and cultural events, could shape the collective perspective. This is referred to as generational awareness. Described by the U.S. Office of Personnel Management (OPM)—a U.S. federal agency responsible for overseeing the federal government's human resources policies and practices to ensure they are effective, equitable, and aligned with government objectives—as the understanding of how significant political, economic, social, and technological events during a certain time period lead to common traits within a generation, in essence, it is about understanding how shared experiences and perspectives can shape each generation's outlook and behavior (U.S. Office of Personnel Management, n.d.).

Generational awareness can provide a framework for identifying traits, preferences, values, and attributes shared within a generation that can then be used to strategically and effectively respond to the specific needs of individuals from all generations. This essentially promotes an inclusive culture where individual differences are considered and valued to more effectively recruit, engage, retain, and develop employees. Gary stressed the importance of ergonomics practitioners considering generational differences to create safer and more efficient work environments for everyone, and that our work of recognizing and addressing these differences should be part of broader Diversity, Equity, and Inclusion (DEI) initiatives.

While the OPM's framework is geared toward attracting, engaging, and retaining generationally diverse employees, the principles are sound advice for us as ergonomics practitioners. The following list is a collection of recommendations derived for each activity based on generational preferences (edited for our purposes as ergonomics practitioners):

Attracting/Recruiting

- Meet people where they "are"—don't rely on one method, such as only reaching out on various social media platforms (I'm not even going to touch the stereotypes around that!); include in-person efforts as well, as that is what Boomers and Traditionalists tend to prefer.
- Ensure that the vision and mission statements resonate with what the various generations can identify with.
- Be flexible in terms of how work gets done, as well as benefits, to attract those with varying needs.
- Avoid snap judgments based on biased stereotypes, especially when determining a person's fit for a position or task.

Onboarding/Training

- Consider how a person would best receive the onboarding experience, such as:
  - Be sure to recognize at the beginning the attributes the person will bring to the organization; for example, depending on the person and their past experience, be sure to immediately recognize and lean on their past experience and expertise, particularly those from a Traditionalist and Boomer generation—also be intentional with formal introductions.
  - For Generation X, be clear and concise with expectations for work, as we (yes, I'm part of the Gen X generation) tend to value working independently.
  - And for Millennials, and now Gen Z, the onboarding experience should emphasize basic expectations for work, as for many this will be their first entry into the full-time workforce.

Engaging Employees

- Communication is key. To meet the various levels of how generations/individuals prefer to be informed, consider using a variety of methods. For example, for critical, wide-reaching information such as organizational changes, consider:
  - For Millennials/Gen Z, set up face-to-face meetings to share how the changes would personally impact them.
  - For Gen X: Share vital information with them immediately in the form of straightforward facts.
  - For Baby Boomers and Traditionalists: Allow them to share past experiences on how they handled similar changes.
- Also consider using a variety of mediums, such as short electronic bursts (via email or other communications platforms), visuals, and in-person meetings (depending on the level of the information).
- In terms of communication and feedback on work instructions and performance, generational suggestions include:
  - For Millennials/Gen Z: They prefer frequent, collaborative feedback in a variety of ways, including quick messaging such as texts, chats, etc.; inject humor.
  - For Gen X: Prefers to be given structure and then left to be self-sufficient; provides immediate blunt feedback on performance via emails or phone calls.
  - For Baby Boomers: Prefer peer-to-peer coaching; prefer important feedback via in-person or phone calls; prefer a friendly rapport and personal touches.
  - For Traditionalists: Prefer informal coaching and recognition of their experience and expertise; prefer information to be shared if and when necessary via personal notes or in-person meetings; prefer formal conversations.

Keep in mind these recommendations are based on the general understanding of how the various generations may perceive them, yet people are individuals. It is always a good idea to have open communication so that feedback can be obtained to better improve the work experience, regardless of the person's generation.

OLDER WORKERS

While each generation comes with its own set of preferences and characteristics, there are also some fundamental truths about different age groups that are particularly relevant to ergonomics. Gary points out that older employees, despite experiencing physical changes with age, bring significant advantages that can greatly benefit ergonomics practitioners. Gary explained,

> …although there are physical changes due to getting older that impact ergonomics design considerations, there are also advantages that older employees have that can be useful to practitioners focused on ergonomics improvements.

This extensive experience can enhance ergonomics processes through the incorporation of a broader range of perspectives. However, these benefits are not always recognized, especially when it comes to older generations. For example, a 2021 AARP survey revealed that over 75% of older workers have encountered age discrimination at work, the highest level since AARP began tracking the issue (Perron, 2021).

Given the current demographic changes discussed in Chapter 6, this issue is increasingly relevant. According to a 2023 Pew Research article (Fry & Braga, 2023), there were approximately 11 million workers aged 65 and older in the U.S. workforce—a number that has nearly quadrupled since the mid-1980s. This increase is largely due to Baby Boomers reaching retirement age, along with changes in retirement plans, such as raising the age eligibility for Social Security Insurance, and the reduction of physically demanding jobs seen with the technology advancements of Industry 4.0. Additionally, improved health and longer lifespans are contributing factors. The U.S. Bureau of Labor Statistics projects that the number of older workers will continue to rise over the next decade (Joint Economic Committee, 2022).

For ergonomics practitioners, it is crucial to consider these factors when designing work environments. By accommodating the needs of older workers, we not only support their well-being but also enhance overall industry performance. And recognizing and leveraging the strengths of all generations, particularly those of older employees, can lead to more effective and inclusive ergonomics solutions.

Designing for an Older Working Population

What are the best practices for designing workspaces for an older workforce? To get some insights, I spoke with Scott Weede, a seasoned Industrial Ergonomics Professional (CIE) with over 25 years of experience in the field. To Scott, DEI is all about fostering a workplace that values and embraces the unique attributes, experiences, and viewpoints of every individual. DEI efforts in the workplace, which he feels ergonomics practitioners such as himself play a role in, help to ensure that all employees receive equal opportunities and feel a sense of belonging.

In recent years, Scott has developed a keen interest in addressing the ergonomics needs of the older working population. He emphasized that the first step in designing for this demographic is to thoroughly understand their distinct characteristics, preferences, and challenges through research and analysis. According to Scott, this foundational knowledge is crucial for making informed design decisions that meet their specific needs,

> Understanding these characteristics will guide your design choices and help you tailor solutions to effectively address the unique needs of older employees.

So, what does this entail?

It starts with generational awareness—familiarize yourself with the typical work habits, technology use, and lifestyle factors associated with different generations. By gaining insight into these aspects, you can create workspaces and tasks that cater to the specific needs, preferences, and limitations of workers from various age groups, including the older worker cohort. This would mean understanding how different generations adapt to technology, their communication preferences and work styles (recommendations from the OPM as discussed earlier), and the physical and psychological challenges various generations may face.

Given the anticipated rise in cyber-physical workspaces and work activities driven by Industry 4.0/Industry 5.0, focusing on the technological preferences and skills of each generation is a critical initial step. Understanding these preferences (for each generation within your workforce) will help you guide the organization in developing strategies to provide appropriate training and support to help bridge any gaps, thus ensuring that all workers can effectively use and benefit from the technologies in the workplace. Along those same lines, it is important to recognize and adapt to the varied communication styles and work preferences of different generations, as knowledge sharing will evolve with these new workspaces. For older generations specifically, such as Baby Boomers and Gen Xers (as much as it pained me to type that), more comprehensive education and training on digital literacy to enhance adoption of new technologies may be warranted.

In addition to general preferences that can enhance the effectiveness of technology and communication, there are specific physiological and psychological changes that vary across generations, particularly associated with older workers. These age-related changes, such as declines in vision, hearing, and mobility, can significantly influence how individuals interact with their work and work environment. Recognizing these changes is essential for designing solutions that minimize—or ideally prevent—challenges that could affect employee safety (including psychological safety) and operational efficiency. Such physical challenges affecting older workers in the context of ergonomics were explored in a study by Vujica Herzog et al. (2023). Their research identified several key issues, including reduced mobility, strength, and balance, as well as diminished eyesight and hearing. Additionally, older workers experienced slower reaction times and decreased dexterity, lower oxygen uptake capacity, and increased systemic blood pressure. The study also noted an accelerated onset of fatigue and heightened sensitivity to extreme temperatures.

Scott emphasized that recognizing and mitigating these physiological changes is only part of the equation. Comprehensive risk assessments must also consider cognitive factors. For aging workers, this includes understanding how cognitive load, memory, attention, and decision-making processes can be affected and what impact they would have on the risk potential of work tasks. For example, research has indicated that in addition to reduced manual dexterity and motor coordination, aging can also lead to reduced numerical and spatial reasoning (Avolio & Waldman, 1990). However, these declines were less pronounced when factors such as experience, education, and job type were considered (Avolio & Waldman, 1994).

A bigger challenge may be with respect to how older workers are perceived by others or how they themselves perceive work and their work environment. While studies show that age has minimal impact on core task performance (Ng and Feldman, 2008), unfortunately, this generation can be perceived as possibly lacking in innovation, performing inadequately, adhering strictly to norms and standards, and being resistant to change (Marchiondo, 2019). This can lead to bias and discrimination, adding to their stress and cognitive strain, as well as them exhibiting higher levels of depression and decreased job satisfaction, which can adversely impact their health.

To address these challenges, consider the following general interventions to address the increased physical and psychosocial risks that could be associated with this generation.

- Utilize adjustable lighting and/or supply task lighting so that workers can increase or lower lighting as needed to enhance visibility, accommodate changes in vision (as well as reduce eye strain).
- Utilize workstations with adjustability options (and encourage use), such as adjustable desks, monitors, and seating to ensure proper posture (neutral posture) and to allow for changes in posture throughout the day.
- Incorporate supportive equipment to aid in physical movement, such as strategically placed handles/grab bars installed in key areas to assist with stability and mobility (e.g., grab bars or cut out to aid in getting in and out of truck cabs; chairs with adjustable armrests to aid in sitting down and getting up).
- Eliminate or minimize excessive force in tasks or integrate mechanical aids to assist with handling and moving objects (e.g., vacuum lifts; a diverter on a conveyor to divert objects closer).
- Design the placement of controls/buttons with sufficient spacing to prevent unintentional activation due to mobility impairments (e.g., tremors or arthritis).
- Ensure controls, labels, etc. are highly visible and readable (considering font size, font type, and contrast ratios) to accommodate reduced visual acuity.
- Allow volume to be adjustable; avoid frequencies beyond 4,000 Hz (between 500 and 1,000 Hz for safety messages) (see Chapter 12 for additional information regarding communications specifically for hearing impairments).
- Promote flexible work policies, such as assisting in designing work that can be done via flexible work arrangements (remote work, flexible working hours, etc.).
- Promote inclusiveness through leadership practices, such as training for managers (including yourself) in recognizing age bias, ensuring generational diversity within your ergonomics teams/participatory activities; consider establishing/promoting mentorship programs so that workers across generations can share knowledge and skills (more details on mentoring are shared later in this chapter).

While it would be a significant challenge to address every potential recommendation for designing work for the older population, additional reference materials are provided at the end of this chapter to support your further research. Although these recommendations focus on the challenges faced by older generations, they are relevant for all age groups, demonstrating how designing for one group's needs can benefit everyone. However, Scott acknowledges that each individual is unique, as are their needs, and a one-size-fits-all solution is often impractical.

> Prioritize allowing individuals to customize their workspace to their comfort and preferences.

Scott concluded by emphasizing that flexibility and adjustability are essential components of any ergonomics design, be it a solution or a program approach, and involving workers in a participatory, engaged process is key.

DESIGNING FOR A YOUNGER POPULATION

When we talk about ergonomics and varying generations, it is easy to focus primarily on the challenges faced by older workers. This is certainly a significant concern, as we have discussed. However, it is equally important not to overlook the physical and social differences that can affect younger employees. These younger workers bring their own set of needs and preferences that can present unique design challenges (or, in a half-full perspective, unique opportunities for innovation).

Gary observed that although younger workers generally possess greater physical strength compared to older workers, these younger workers are often more susceptible to workplace injuries. He posited that it could be due to their tendency to engage in risk-taking behaviors and their relative inexperience. He suggested that these workers may not yet have mastered safe work practices or developed the level of caution typically observed in their more seasoned counterparts, leading to an increased risk of injury.

Gary also highlighted a notable challenge in managing generational differences in the workplace, specifically as related to younger workers: their reduced willingness to accept and endure physically strenuous tasks. In a 2017 article discussing the challenge of attracting the next generation of skilled construction labor, 63% of the undecided young adults (18- to 25-year-olds) surveyed about career choices indicated there was no or little chance they would consider a career in the trades, with 48% simply saying they wanted (what they perceive as) a less physically demanding job (McManus, 2017).

This can be attributed to a few factors. One key factor is the prevailing preference among workers—particularly the younger demographic—for careers that offer a greater sense of purpose. According to a survey by McKinsey & Company, 70% of American workers expressed a desire for careers with a meaningful impact, and physically demanding jobs are often not viewed as fulfilling this criterion (Horovitz, 2022). Another reason is the demographic shift, particularly in the U.S., with workers having more advanced education and shifting to work that is more reliant on advanced technology as opposed to less-skilled physical labor. As a sidenote: despite the underlying reasons for why the younger generation is not filling these physically demanding jobs, such as those seen in construction, agriculture, and hospitality, the need to fill them remains high, leading to another diverse workforce challenge for ergonomics as seen with the increase of foreign-born workers taking on these high physical demand jobs.

Gary further explained that the younger generation's hesitation to take on physically demanding roles can contribute to an increased risk of injury due to limited exposure. The risk in these jobs is amplified when these workers have not yet developed effective strategies for managing and mitigating potential hazards. Their inexperience and lack of preparation in handling strenuous tasks can make younger employees more susceptible to workplace accidents, as they may not yet possess the skills or knowledge needed to navigate and reduce risks effectively.

Child Labor

While there are global and regional conventions and laws aimed at ensuring safe working conditions for children, such as the International Labour Organization Minimum Age Convention, the E.U.'s Directive on Child Labor, and the U.S.'s Fair Labor Standards Act, it is not uncommon for children to be found working in conditions detrimental to their physical and mental health. Such was shared in an article by Human Rights Watch's Margaret Wurth, then a Senior Researcher for their Children's Rights Division. In that article, Ms Wurth discussed interviewing hundreds of child workers working in unsafe conditions, such as long working hours in extreme heat or working with heavy machinery. According to a 2021 UNICEF report, roughly 1 in 10 of all children worldwide (between the ages of 5 and 17) are engaged in child labor (UNICEF, 2021). As the report further noted, this represented a stall in the effort to end the practice of child labor. Additionally, recent political climates dictating changes in laws and ordinances, particularly in the U.S., have further increased the potential for the number of younger workers in the workforce. The reality of child labor statistics and the increasing shift of more younger workers in the workplace make it crucial for ergonomics practitioners to be aware of and be prepared to address these emerging challenges. For instance, younger workers may have different physical demands, technology habits, and work style preferences that could impact how they interact with their work environment. And we need to be proactive in understanding and addressing these needs.

For starters, the United States' Occupational Safety and Health Administration has shared several concerns when it comes to the safety of younger workers, such as potential inadequate training on the use of equipment or processes, inadequate supervision, or youth performing dangerous work deemed illegal or inappropriate for them. According to a 2018 U.S. Government Accountability Office report, more U.S. child workers die in agriculture than in any other industry due to working conditions such as exposure to excessive heat (GAO, 2018). Beyond death due to unsuitable conditions, the 2020 CDC Report assessing Nonfatal Occupational Injuries to Younger Workers in the United States (2012–2018) indicated that younger workers, aged 15–25 years and who, during the time of the report, made up 13% of the working population in the U.S., were more likely to be associated with workplace hazards like violations of child labor laws, fast pace of work, minority status, and lack of skills, experience, supervision, and high-quality safety training (CDC, 2020).

Furthermore, the report noted that young workers might be less likely to recognize workplace hazards, voice safety concerns, or be aware of their legal protections.

As ergonomics practitioners, we need to recognize that there can be unique risks and hazards for this group of workers as part of a diverse workforce. As with older workers, where physical, cognitive, and social differences need to be accounted for, so should our focus be on the physical, cognitive, and social differences of younger workers. For example, digital eye strain and dry eyes have been recognized as issues for the younger population in Kaur et al. (2022)'s comprehensive review, noting that cohort's propensity for consuming digital media, which was exacerbated by the COVID-19 pandemic. And in Landau et al. (2008)'s study, they found that younger workers were associated with unfavorable higher workload jobs, which could increase the risk of injury.

Beyond age differences, according to Benjamin and Wilson (2005), "lifestyle, education, socio-economic status, genetics, stress, exercise, nutrition, and healthcare needs may have an equal if not greater importance" and therefore should be stressed just as much to younger workers. In fact, education on risks in the workplace is an important key in helping to manage risks. The National Institute for Occupational Safety and Health, recognizing the importance of educating younger workers, particularly those 18 and younger, has developed a specific training program called "Youth@Work—Talking Safety" (CDC, n.d.) which is dedicated and aimed at providing guidance to schools to help educate children who may venture out into the workforce. This includes guidance on safety hazards, including physical hazards such as heavy lifting, awkward postures, and repetitive motions, as well as information specific to state laws governing child labor.

THE IMPORTANCE OF STAYING INFORMED

Considering the various factors in a diverse workforce, including generational differences, both Scott and Gary highlighted the crucial need for ergonomics practitioners—and students in the field, as noted by Gary—to remain updated on the latest research, trends, and best practices, particularly those pertaining to different generations. A comprehensive awareness and understanding of these elements are essential to effectively design for diversity, and staying informed is a vital component of any ergonomics practitioner's toolkit.

There is a wealth of research available on designing work environments for older populations, including safety considerations (Czaja & Sharit, 2023), the specific needs of older construction workers (Ranasinghe et al., 2023), and the impact of technology on older workers' stress levels (Nedeljko et al., 2024). However, as noted by Bentley et al. (2023), more research is needed to address gaps in occupational health and safety interventions for older workers. Their review highlights the need for future studies focusing on technological advancements, evolving work-life expectations, and policy changes across different regions and industries, considering both physical and psychosocial aspects.

Scott stressed the critical need for continuous education and training, both within organizations and through professional associations. It goes without saying that staying up-to-date with the latest research and best practices is essential for effectively applying ergonomics principles. Ongoing learning helps ensure that our approaches remain relevant and effective as the field evolves. But fostering collaboration and knowledge sharing across different generations is just as crucial. Younger professionals often bring new perspectives and technological expertise, which can drive innovation. Meanwhile, more seasoned colleagues provide valuable insights based on their extensive experience and historical context.

Professional associations and societies, like the Institute of Industrial and Systems Engineering (IISE)'s Applied Ergonomics Society (where I had the honor of serving as President) and the Human Factors and Ergonomics Society, play a vital role in keeping ergonomics practitioners updated on these evolving topics. They provide numerous ways to stay informed, including workshops and webinars that explore new ideas and share practical advice. Professional conferences offer another excellent opportunity to hear from experts, engage in discussions, and network with peers. Additionally, periodicals, newsletters, and blogs offer quick updates and insights. Perhaps the most dynamic way to stay engaged is through real-time discussions and interactive exchanges, where members and guests can network and share information, discuss examples and case studies, and learn from each other's experiences to understand how cutting-edge research is applied in real-world scenarios.

When it comes to understanding work trends, Glassdoor's annual Workplace Trends reports (Glassdoor, 2024) is a good resource. This report taps into insights from millions of professionals who engage with Glassdoor each month, covering varied topics from generational trends in the workplace to salary and benefit data to remote work policies to other factors that shape the modern work environment.

Staying informed about trends is important, but it is only part of the equation. To effectively engage all demographics, particularly in ergonomics, adopting an iterative, user-centered design approach is essential. This involves continuously gathering feedback, pinpointing issues, developing solutions, and fine-tuning strategies. I have emphasized the value of this approach throughout this book—how it should be integrated into both the design and implementation of ergonomics solutions. For instance, by conducting direct observations, interviews, and usability tests with different age groups, we can better tailor our solutions to their unique needs. By integrating these diverse viewpoints, we can tackle workplace challenges with a blend of cutting-edge ideas and tried-and-true methods. This approach not only addresses current issues but also demonstrates our commitment to continuous improvement in ergonomics. It ensures that we are leveraging the full spectrum of knowledge and experience available, ultimately benefiting everyone involved and advancing the field.

APPRECIATING GENERATIONAL DIFFERENCES IN THE WORKPLACE

Scott highlighted that while significant strides have been made in moving away from broad generalizations and stereotypes about older workers, there is still considerable work to be done in truly understanding and valuing the individuality and unique experiences of each generation, emphasizing.

> …there is still a need for greater recognition and awareness regarding the conversation around generational differences.

This ongoing challenge highlights the critical need to address the diverse needs and perspectives of the various generations in the workplace, particularly the older generations, rather than relying on broad and often inaccurate assumptions. To effectively tackle this issue, mentoring can be a powerful tool to bridge generational gaps, promote mutual respect, and create a more inclusive and productive work culture.

The Importance of Mentoring

Mentoring programs are an excellent way for organizations to foster generational awareness and promote knowledge sharing. They not only help deepen understanding between different age groups but also enhance overall collaboration within the workforce. Returning to the guidance provided by the OPM, mentoring typically involves a formal or informal relationship between a more experienced individual (the mentor) and a less experienced one (the mentee) (U.S. Office of Personnel Management, n.d.). This relationship can be between an older or senior colleague and a junior colleague, though it is not always strictly defined by age or seniority.

Mentoring allows employees from various generations to exchange unique experiences and insights, leading to more nuanced and effective solutions. The mentoring relationship is designed to focus on productive professional development in specific areas, which are often discussed and agreed upon by both parties involved. While mentees benefit from guidance and career growth, mentors also gain from the experience. They develop leadership skills and soft skills such as empowerment, conflict resolution, and time management. Additionally, mentors enjoy the respect and recognition that comes from being seen as valued experts and teachers for the next generation.

Beyond the personal development of both mentors and mentees, organizations benefit from improved employee retention, positive culture changes, and increased generational awareness. Therefore, ergonomics practitioners should actively seek out mentoring opportunities, whether through formal programs within their organizations or by volunteering with professional ergonomics and human factors societies.

I would be remiss if I didn't highlight my own experiences with mentoring, both as a mentee and a mentor. My current career—or rather, my profession—owes a great deal to a fantastic mentor who guided me early on. When I first started in ergonomics, I was fortunate to be mentored by Marenda, who has also played a significant role in this book's creation. Through Marenda's guidance, I honed my skills as both a practitioner and a leader, and I can look back over the past 20 years and see how far I've come in my profession. It is a feeling every ergonomics practitioner, regardless of their age or experience, should experience.

With this in mind, I have made it a priority to mentor others, some of whom have contributed to this book. One standout example is LaShawn Boulware, a CPE with 10 years of experience who was a Human Factors Engineer working in the aerospace industry. LaShawn has transitioned her skills into business development, communication, and career coaching, and she's become a sought-after speaker on networking and mentoring. From her own mentorship journey, LaShawn emphasizes that whether you're mentoring or being mentored, both parties gain valuable insights from their different perspectives. LaShawn points out that learning flows both ways in mentorship relationships. She noted,

> Whether I'm mentoring or being mentored, regardless of age, we both benefit from each other's viewpoints.

Additionally, LaShawn highlights the value of diverse mentorship relationships. Her peer mentor provides a space for sharing experiences that she might not discuss with a senior mentor. Beyond traditional mentoring, LaShawn has also engaged in career coaching mentorship, which introduced her to an internal coaching program. This opportunity allowed her to coach high-performing employees within her previous company and significantly accelerated her career growth.

> Whenever someone reaches out to me for mentorship, whether formal or informal, I'm eager to offer my insights. One mentee, after hearing me speak at an award ceremony, reached out because she wanted to learn from my experience with NASA astronauts [as part of her company's highly touted human factors design program]. This mentorship has been a rewarding exchange of knowledge and personal growth for both of us.

For ergonomics practitioners, LaShawn advised that building and maintaining mentorship relationships demands time and effort. She quoted, "*Create your tribe of believers who want to imagine and manifest the impossible with you*," reflecting her approach to nurturing trust and mutual growth. Learning from mentors of all ages provides critical support and alignment for achieving success. She emphasized that while you may not be an expert in every area, it is essential to find those who are. Whether through work colleagues, LinkedIn connections, or family members, the key is to be intentional in seeking guidance.

So, if you haven't yet engaged in some form of mentorship, make it a priority. The profession thrives on mutual support and collaboration, and taking care of one another is vital to our collective success.

FURTHER READING

For additional reading on designing and mitigating for challenges related to varying generations, consider the following:

- Designing For Older Adults: Principles and Creative Human Factors Approaches (Fisk et al., 2004).
- No time to retire: Redesigning work for our aging workforce (Schwartz et al., 2018).
- Ergonomics for the inclusion of older workers in the knowledge workforce and a guidance tool for designers (Gonzalez & Morer, 2016).
- IISE presentation—The Aging Workforce and Impacts on Ergonomic Design (Institute of Industrial and Systems Engineers [IISE], n.d.).

- Ergonomics/Human Factors Needs of an Aging Workforce in the Manufacturing Sector (Stedman et al., 2012).
- Ergo+50: Ergonomic Assessment Methodology Aimed at Older Workers (Remesal et al., 2020).

REFERENCES

Avolio, B. J., & Waldman, D. A. (1990). An examination of age and cognitive test performance across job complexity and occupational types. *Journal of Applied Psychology*, *75*(1), 43. https://doi.org/10.1037/0021-9010.75.1.43

Avolio, B. J., & Waldman, D. A. (1994). Variations in cognitive, perceptual, and psychomotor abilities across the working life span: Examining the effects of race, sex, experience, education, and occupational type. *Psychology and Aging*, *9*(3), 430. https://doi.org/10.1037/0882-7974.9.3.430

Benjamin, K., & Wilson, S. (2005). *Facts and misconceptions about age, health status and employability* (Report No. HSL/2005/20). Buxton: Health and Safety Laboratory.

Bentley, T., Onnis, L. A., Vassiley, A., Farr-Wharton, B., Caponecchia, C., Andrew, C., & Green, N. (2023). A systematic review of literature on occupational health and safety interventions for older workers. *Ergonomics*, *66*(12), 1968–1983. https://doi.org/10.1080/00140139.2023.2221332

Centers for Disease Control and Prevention. (2020). Workplace safety and health: Employment and training programs could improve support for workers with disabilities (MMWR, Vol. 69, No. 35). https://www.cdc.gov/mmwr/volumes/69/wr/pdfs/mm6935a3-H.pdf

Centers for Disease Control and Prevention. (n.d.). *Talking safety*. National Institute for Occupational Safety and Health. Retrieved November 11, 2024, from https://www.cdc.gov/niosh/talkingsafety/default.html

Czaja, S. J., & Sharit, J. (2023). Safety considerations for older workers. In *A comprehensive guide to safety and aging* (pp. 179–186).

Fisk, A. D., Rogers, W. A., Charness, N., Czaja, S. J., & Sharit, J. (2004). *Designing for older adults: Principles and creative human factors approaches*. CRC Press/Taylor & Francis.

Fry, R., & Braga, D. (2023, December 14). Older workers and the changing workplace. *Pew Research Center*. https://www.pewresearch.org/wp-content/uploads/sites/20/2023/12/ST_2023.12.14_Older-Workers_Report.pdf

Gonzalez, I., & Morer, P. (2016). Ergonomics for the inclusion of older workers in the knowledge workforce and a guidance tool for designers. *Applied Ergonomics*, *53*(A), 131–142. https://doi.org/10.1016/j.apergo.2015.09.002

Glassdoor. (2024). Workplace trends 2024. *Glassdoor*. https://www.glassdoor.com/research/workplace-trends-2024

Horovitz, B. (2022, May 13). *Employers take note: Young workers are seeking jobs with a higher purpose. TIME*. https://time.com/6176169/what-young-workers-want-in-jobs/

Institute of Industrial and Systems Engineers (IISE). (n.d.). *The Aging Workforce and Impacts on Ergonomic Design. IISE*. Retrieved Oct 9, 2024 from https://www.iise.org/details.aspx?id=41882

Joint Economic Committee. (2022, December 21). *The number of older workers is growing quickly as they continue to face economic challenges*. United States Senate. https://www.jec.senate.gov/public/index.cfm/democrats/2022/12/the-number-of-older-workers-is-growing-quickly-as-they-continue-to-face-economic-challenges

Kaur, K., Gurnani, B., Nayak, S., Deori, N., Kaur, S., Jethani, J., & Mishra, D. (2022). Digital eye strain—a comprehensive review. *Ophthalmology and Therapy*, *11*(5), 1655–1680. https://doi.org/10.1007/s40123-022-00477-x

Landau, K., Rademacher, H., Meschke, H., Winter, G., Schaub, K., Grasmueck, M., et al. (2008). Musculoskeletal disorders in assembly jobs in the automotive industry with special reference to age management aspects. *International Journal of Industrial Ergonomics*, *38*(7–8), 561–576. https://doi.org/10.1016/j.ergon.2007.10.007

Marchiondo, L. A., Gonzales, E., & Williams, L. J. (2019). Trajectories of perceived workplace age discrimination and long-term associations with mental, self-rated, and occupational health. *The Journals of Gerontology: Series B*, *74*(4), 655–663. https://doi.org/10.1093/geronb/gbx095

McManus, J. (2017, April 20). Why don't young Americans want to do construction work? *Builder Online*. https://www.builderonline.com/building/trades-subcontractors/why-dont-young-americans-want-to-do-construction-work_o

Nedeljko, M., Gu, Y., & Bostan, C. M. (2024). The dual impact of technological tools on health and technostress among older workers: An integrative literature review. *Cognition, Technology & Work*, *26*(1), 47-61. https://doi.org/10.1007/s10111-023-00699-7

Ng, T. W. H., & Feldman, D. C. (2008). The relationship of age to ten dimensions of job performance. *Journal of Applied Psychology*, *93*(2), 392–423.

Perron, R. (2021). Understanding older workers during the coronavirus pandemic. *AARP Research*. https://doi.org/10.26419/res.00445.001

Ranasinghe, U., Tang, L. M., Harris, C., Li, W., Montayre, J., de Almeida Neto, A., & Antoniu, M. (2023). A systematic review on workplace health and safety of ageing construction workers. *Safety Science*, *167*, 106276. https://doi.org/10.1016/j.ssci.2023.106276

Remesal, A. F., Goossens, R., & Murata, A. (2020). Ergo + 50: Ergonomic assessment methodology aimed at older workers. In R. Goossens & A. Murata (Eds.), *Advances in Social and Occupational Ergonomics* (Vol. 970, pp. 35–46). *Advances in Intelligent Systems and Computing*. Springer. https://doi.org/10.1007/978-3-030-20145-6_3

Schwartz, J., Monohan, K., & Anderson, S. (2018, December 7). Redesigning work for our aging workforce. *Deloitte Insights*. https://www2.deloitte.com/us/en/insights/focus/technology-and-the-future-of-work/redesigning-work-for-our-aging-workforce.html

Stedmon, W. A., Howells, H., Wilson, R., & Dianat, I. (2012). Ergonomics/Human factors needs of an ageing workforce in the manufacturing sector. *Health Promotion Perspectives*, *2*(2), 112–125. https://doi.org/10.5681/hpp.2012.014

U.S. Government Accountability Office (GAO). (2018). Workplace safety and health: Employment and training programs could improve support for workers with disabilities (Report No. GAO-19-26). U.S. Government Accountability Office. https://www.gao.gov/products/gao-19-26

U.S. Office of Personnel Management (OPM). (n.d.). Leading a generationally diverse workforce. https://www.opm.gov/policy-data-oversight/training-and-development/reference-materials/online-courses/leading-a-generationally-diverse-workforce/LeadingAGenerationallyDiverseWorkforce.pdf

U.S. Office of Personnel Management (OPM). (n.d.). Mentoring. U.S. Office of Personnel Management. https://www.opm.gov/policy-data-oversight/training-and-development/mentoring/

UNICEF. (2021, June 9). Child labour 2020: Global estimates, trends, and the road forward. *UNICEF*. https://data.unicef.org/resources/child-labour-2020-global-estimates-trends-and-the-road-forward/

Vujica Herzog, N., Kaya, O., & Buchmeister, B. (2023). Ergonomics and ageing – Case study. In B. Katalinic (Ed.), *DAAAM International Scientific Book 2023* (Vol. 15, pp. 017-032). DAAAM International. https://doi.org/10.2507/daaam.scibook.2023.02

11 Inclusive and Equitable Design

What does ergonomics have to do with DEI?

This was an off-hand question I heard in a conversation about Diversity, Equity, and Inclusion (DEI). It happened a while ago; in fact, I think it sparked my journey to show the connection between ergonomics and DEI. And my response to that question?

Everything!

I don't share this question in judgment, nor do I fault the person who asked it, as it was a genuine inquiry (and likely one that you or someone you know in this field may have had or may still have). From what I recall, the context of the question was around how ergonomics could play a role in DEI, particularly when "DEI is about racial diversity." Right away, it is clear that this person's view of "DEI" was somewhat limited, as we know that while racial diversity is crucial to the conversation around diversity, equity, and inclusion, DEI encompasses much more than just race or ethnicity. Perhaps less obvious, though, was their limited view of ergonomics, as ergonomics actually has a significant role to play in fostering racial, cultural, ethnic, and other forms of identity diversity. If you have made it this far in the book, I hope you're starting to see the connection and potentiality.

In previous discussions, we've seen how ergonomics clearly influences work design when physical differences are apparent. This was especially evident in earlier chapters where physical considerations were front and center. However, designing for diversity goes beyond addressing visible physical differences and extends into the realms (and intersections) of racial, cultural, LGBT+, and socioeconomic diversity.

In my conversation with Gary (Allread), who shared his perspectives on designing for varying generations, he noted how psychosocial factors, such as stress and job satisfaction, are crucial to consider in ergonomics and ergonomics processes, alongside physical ergonomics. He pointed out that incorporating diverse perspectives, including those of LGBTQ+ and racial/ethnic minorities, is essential for creating effective solutions, though these factors are more psychosocial than physical. Understanding how personal history or experiences influence employees' interpretations and responses to work is vital. And studies have shown there are significant links.

Perceived workplace discrimination is linked to poorer health and well-being, as individuals who face discrimination must constantly manage negative experiences, increasing their cognitive load (Xu & Chopik, 2020). This vigilance against biases and microaggressions can lead to chronic stress and anxiety, depleting cognitive resources and impairing focus, information processing, and decision-making

 DOI: 10.1201/9781032658988-14

(Mottaghi et al., 2024). Over time, these stressors can result in burnout, decreased productivity, and increased errors (Mottaghi et al., 2024). Research has indicated that microaggressions and instances of discrimination significantly affect mental well-being and cognitive demands (Brar et al., 2019). Additionally, stressors that exceed an individual's ability to cope can trigger detrimental physiological reactions, especially in older individuals.

While inclusiveness is crucial, Gary felt the focus should be on fostering a supportive environment just as much as on physical design aspects. Physical aspects certainly impact the effectiveness of designs; the real challenge—and what this chapter will focus on—is the failure to consider these diverse dimensions. When designs do not account for the broad spectrum of diversity, they can inadvertently exclude certain populations, leading to significant issues of ineffectiveness and marginalization.

This chapter will examine how neglecting these aspects of diversity can result in designs that fail to meet the needs of all users. Additionally, it will further explore the psychological stress associated with exclusion, a key concern for ergonomics practitioners as understanding and addressing these factors are imperative for creating inclusive environments that support everyone effectively and equitably.

DESIGNING FOR RACIAL DIVERSITY

Designing for inclusion, especially as related to job accommodations in particular, was significantly influenced by the civil rights movement in the U.S., which advocated for the equal rights and opportunities of various oppressed groups, including racial minorities, women, and LGBTQ+ individuals. Embracing racial and cultural diversity in the design of work ensures that products, processes, policies, and practices meet the needs of a varied population, promoting equal opportunities and participation for everyone. This approach not only drives innovation by incorporating diverse perspectives and addressing a wider range of needs, but it also enhances fairness, which the perception of and its impact on individuals' overall well-being will be discussed in more detail as we go through this chapter.

From a racial and ethnic diversity perspective, taking an inclusive design approach means taking into account not only physical differences—such as skin color, facial features or body anthropometry—that can dictate design characteristics, but also the experiences, values, and expectations that may have been shaped due to a person's racial identity. All of these differences should be taken into account for the overall look, feel and perception of the process or product and how the individual plans to interact with it. This approach ensures that the design is responsive to the diverse needs of users and considers how different racial/ethnic backgrounds can affect interaction with the design. An essential aspect of this inclusive design process is actively avoiding stereotypes, biases, and discrimination. It is crucial not only to prevent intentional biases but also to ensure that designs do not inadvertently exclude individuals based on their racial or cultural identity. This means carefully evaluating how design elements might possibly impact different racial groups and striving to create solutions that are equitable and accessible for everyone.

For instance, a notable example of racial diversity considerations in design (or limitation thereof) is the issue with widespread CO_2 oximeters designs, which have been found to be less accurate in measuring blood oxygen levels in individuals with darker skin tones.

Pulse Oximeters and Darker Skin

Chances are, when you visit your doctor or general practitioner, part of the standard intake process will include measuring your blood pressure, temperature, and pulse. This routine involves using equipment designed to provide quick and accurate readings of your vital signs and are necessary for assessing your current health status and determining any necessary treatment. However, with devices like pulse oximeters, "accuracy" was not always guaranteed, particularly for individuals with darker skin color, which can have serious implications for patient care.

Pulse oximeters are medical devices used to measure the oxygen saturation levels in the blood. Accurate readings are vital because low blood oxygen levels can be life-threatening, necessitating prompt and appropriate medical intervention. During the COVID-19 pandemic, pulse oximeters were particularly valuable for rapidly assessing patients' oxygen levels to guide treatment decisions. Typically, these devices were placed on a patient's finger or earlobe, where they use light sensors to gauge oxygen saturation through the tissue. While pulse oximeters were preferred for their ease of use and widespread availability, they were not as precise as arterial blood-gas tests, which are considered the gold standard but are not always readily accessible.

A study conducted by Sudat et al. (2023) revealed a significant issue with pulse oximeters during the pandemic: they were less accurate for Black patients due to their darker skin tones. This inaccuracy led to an overestimation of oxygen levels, potentially resulting in delayed or inadequate treatment for these patients. The device's design inadvertently introduced racial bias, causing disparities in care by failing to account for variations in skin pigmentation. This case illustrates how reliance on biased technology can exclude certain populations from receiving timely and appropriate medical attention.

The study recommended several critical changes to address these disparities. It emphasized the need for healthcare professionals to question pulse oximeter readings when other clinical indicators do not align with the device's measurements. Additionally, it advocated revising the testing guidelines for pulse oximeters and similar devices to ensure they reflect a more diverse patient population. The oversight of including a diverse range of skin tones in the device's accuracy testing led to a significant delay in identifying and addressing the issue, highlighting the need for inclusive design practices to prevent such biases from compromising patient care.

Racial Bias and Inequity in Technology

For those of us that have set up our phones to recognize our faces to unlock them, we probably don't think about how it works. In fact, facial recognition is so ubiquitous these days that we may take the technology for granted. That is, until it doesn't work for you (how many of you have tried and failed to unlock their phone with half their face buried in a pillow with the intention of scrolling social media in bed. The sheer act of having to raise your head three inches can be tough!). But imagine if that technology, which is used in many applications today in society—such as for social media photo tagging, building access, airport screening, and law enforcement—was the reason you were incorrectly denied access or entry, detained, or imprisoned.

Facial recognition is the use of specific landmarks on an individual's face to analyze and verify a person's identity based on comparisons to previously collected images of the person. Or so that is how it works in theory. Of all the biometrics available today used for identification, facial recognition is the least accurate. And that accuracy can be biased toward individuals with certain racial features, such as skin color, as identified by the National Institute of Standards and Technology (2019) as part of a series of reports analyzing bias in facial recognition programs. Various programs were found to perform differently in the processing of images of individuals in various demographics, particularly for women of color, where it found that 189 algorithms performed least accurately. These errors in detection can lead to negative outcomes for people, such was the case of Randal Quran Reid where a facial recognition program used in law enforcement led to his wrongful arrest (AP, 2023). Unfortunately, Mr Reid's case is not a singular event of civil liberties being violated due to racial bias within facial recognition technology. As such, calls have been made for the technology and the application of the technology to be changed to address the bias, or at the very least, limit the sole reliance of the technology in law enforcement. But what were those suggested changes?

A noted issue leading to potential racial bias in facial recognition technology relates to the data used to train and assess the algorithms, which tends to be predominantly white males according to a study conducted by Buolamwini and Gebru (2018). As such, one ask for developers is to ensure that the datasets, in terms of development as well as sources used for comparison, are more diverse and optimized to minimize potential racial bias. Furthermore, there is an ask for more transparency in not just the development of algorithms, but also the performance of these programs across diverse groups for accountability.

Recent discussions about facial recognition have highlighted issues of racial bias and inequitable treatment. However, the potential for bias—both racial and gender—exists in other technologies that incorporate physical attributes in their design. For example, dispensers that use near-infrared technology to detect skin tones often fail to recognize darker skin, which can lead to significant problems beyond mere inconveniences, similar to the risks associated with flawed facial recognition systems. Even if these design flaws seem minor, their persistent occurrence across various technologies can be disheartening. As ergonomics practitioners, we must remain vigilant about these potential biases in our designs and the tools used within our processes (e.g., risk assessments, anthropometric tools). It is crucial to involve diverse perspectives in the development and validation of our products, processes and tools and to actively ask how we might unintentionally exclude certain groups. Furthermore, we should advocate for education and awareness about diversity, particularly in recognizing the varied ways human bodies and minds function, and ensure these differences are integrated into our designs.

Beyond the concerns with facial recognition technology, the growing use of sensory technologies, like voice command and voice recognition software, could pose significant challenges for individuals with accented speech or those who speak using local dialects. For example, one study (Koenecke et al., 2020) conducted on voice transcribing tools found that the tools misunderstood Black speakers nearly twice as often as it did white speakers.

If these diverse speech patterns are not considered during the design and development of these tools, many users may find them ineffective or frustrating to use—all potentially impacting operational efficiency. It is critical that these tools are developed including diverse linguistic data, which should be done by actively seeking feedback from potential users with various speech patterns.

As ergonomics practitioners, we are working in an era where technological advancements are evolving at an extraordinary pace. This is undoubtedly a positive development, especially when it comes to enhancing our ability to recognize and mitigate risks more effectively. However, as we embrace these innovations, it is crucial that we assess each new technology not only for its effectiveness but also for its relevance and applicability to the diverse workforces we serve. The potential of advanced technology in ergonomics is exciting, but as we incorporate these technologies into our practices and ergonomics processes, we must remain alert about not just ensuring, but demanding as well, that the tools and equipment we rely on are not just effective but also equitable.

A Cautionary Tale

As a profession, we are deeply invested in the user experience. For reasons why it is so important one only has to look at the plethora of lessons learned from the beauty brand Youthforia when it became embroiled in controversy when they attempted to expand their foundation color collection to be more inclusive but responded badly, as many would say, to criticism that they missed the mark (LeSavage, 2024). While you may not have to deal with finding the perfect shade of foundation makeup, the lessons from the Youthforia issue remains—whether the design is a product intended for a wide range of people, or work processes intended for a diverse workforce, it is crucial to understand the execution of inclusive design to minimize unintentional consequences.

From a psychosocial risk perspective, perceptions of exclusion can lead to adverse emotional and behavioral effects, such as emotional discomfort (frustration or anger), a threatened sense of identity worth, evoking a sense of rejection, or the internalization of inferiority. From an ergonomics standpoint, psychological safety plays a vital role in ensuring that employees feel comfortable and empowered to engage with their environment. When workers feel psychologically safe, they're more likely to contribute fully and work collaboratively, which leads to better outcomes in both product design and workplace processes. According to Psychology Today (Edmondson, 2020), measuring psychological safety is not only crucial for tracking DEI progress but also for understanding subjective perceptions of inclusion, which are just as important as objective DEI metrics (See Part III for more information on the various tools used to measure DEI and ergonomics progress).

As discussed in Part I, the user experience is inherently diverse. We're concerned with more than just the physical aspects of a product or process or even the user's health and safety. We also focus on the emotional and cognitive elements, including usability, user satisfaction, and accessibility. The "user" in this context includes people with a wide range of physical and cultural experiences and perspectives.

A positive user experience often leads to increased product usage or purchases, and the same can be said for the products or processes we design. Our goal as ergonomics practitioners is to create something akin to "brand loyalty" for our designs by ensuring they are thoughtfully and inclusively developed. When users—or workers, in the case of workplace design—feel seen, valued, and heard, they become loyal advocates. This expands our reach and impact. To achieve this, we must employ a truly user-centered approach to inclusive design. At the end of the day, inclusive design isn't just a matter of good ethics—it is simply good business. Organizations that prioritize positive user experiences in their designs will find success, and those that take diverse user experiences into account will be poised to thrive.

Practically Designing for Racial and Cultural Diversity

For ergonomics practitioners, when designing with racial and cultural diversity in mind, we need to design work environments, products and processes that are not only functional but are also considerate of the diverse needs and expectations shaped by the individuals' varied racial or cultural backgrounds (or intersection thereof) and life experiences. This calls for being intentional with trying to understand of how cultural, social, and historical context can influence user behavior and interaction with our designs.

In the Opening Introduction, the correlation between ergonomics and the meeting of basic human needs as a means of ensuring a sense of belonging were discussed. In considering the workforce's diversity, it is essential to recognize that the hierarchy of needs will differ across the spectrum of employees. While basic physiological needs may be foundational for all, how that manifests is dependent on the individuals, which are different and are influenced by many factors such as their geographical location, their climate conditions, their culture and customs, societal expectations, etc. Discussed by Flynn, Castellanos, and Flores-Andrade in their article for "Professional Safety" about safety across cultures (2018), these differences, such as varying cultural expectations of gender roles with respect to physical work or perceptions of work ethics, can challenge the effective implementation of policies and procedures and strategies must be designed with cultural norms in mind.

For example, consider fatigue management initiatives, particularly as related to the cultural norm of the siesta. In many warmer climates, a midday nap is a beloved tradition that effectively helps combat fatigue. For an ergonomics practitioner aiming to roll out a comprehensive fatigue management policy across an organization with diverse countries and cultures, it is crucial to recognize that while such a policy might be welcomed in some places, thus improving the potential for effective implementation (including operational accommodations), it could face pushback in other locales.

Additionally, continuing with the cultural fatigue example, in the development of a comprehensive fatigue management initiative, we should also consider how different racial or cultural groups may perceive fatigue (Karasz & McKinley, 2007; Cordero et al., 2012) or how they would perceive any guidance provided with regards to sleep patterns (Sephton & Kay, 2024; Estafanous & Sedky, 2023), shift schedules or program management (Presser, 2003). Or, as another example, in the design of stress management initiatives (in correlation with cross-functional organizational efforts to mitigate identified psychosocial risks), there should be consideration with how various cultures perceive and respond to stress, which can be impacted by communication styles and language used to discuss and educate about stress, as well as cultural stigma surrounding mental health and access to it (Sue & Sue, 2012).

When we're developing ergonomics initiatives, it is essential to acknowledge the cultural nuances that can significantly influence our efforts. Different cultures and communities may respond in varied ways to proposed strategies, so understanding these dynamics is crucial for creating effective solutions. While it would be impossible to capture all potentialities of racial or cultural needs that might require design adjustments, the essential point is to acknowledge that people from diverse backgrounds may have unique experiences, preferences, and expectations that can influence how they interact with work environments, products, or processes. And when we recognize these differences, we can address them, preferably during the design process. But probably most importantly, ensure there is clear communication and engagement with employees from diverse backgrounds, again, preferably during the design phase. This will help to ensure that critical concerns are highlighted and considered, such that the initiative is effective and equitable.

Ensuring that our approaches meet the unique needs and preferences of each group is vital, and it may mean that ergonomics solutions or initiatives implemented in one location may need to be adjusted or adapted as needed to ensure it resonates with the relative cultural diversity of the workforce. This means understanding and being culturally aware of the diverse makeup of the workforce in which you seek to implement ergonomics initiatives and recognizing any potential physical and/or cultural differences that may impact ergonomics efforts. Furthermore, from an organizational ergonomics process development approach, allowing enough freedom in how individual sites within your organization execute the prescriptive ergonomics process ("freedom in the framework") will better allow for this flexibility.

Universal Design and Inclusive Design—And How Each Can Be Applied

As discussed in Chapters 3 and 8, Universal Design principles are crucial for developing products and systems that are accessible and effective for *everyone*, regardless of individual differences. This approach can be especially important when physical attributes are integral to the design's functionality, such as in facial recognition technologies or skin-detection sensors. To achieve truly inclusive and equitable designs, it is essential that these technologies are developed and validated with a comprehensive understanding of the diverse range of racial/ethnic identities and physical characteristics of the users.

Let's take facial recognition programs as an example. These programs must be designed to accurately and reliably identify individuals across various facial features and skin tones. This requires a thorough, representative dataset that includes a broad spectrum of racial/ethnic identities to prevent biases and ensure that the technology performs consistently for all users. Similarly, skin-detection sensors, which might be used for health monitoring or for security purposes, need to be calibrated and tested to account for differences in skin pigmentation and other physical traits to avoid inaccuracies and potential exclusions.

Incorporating Universal Design principles means not only focusing on the physical aspects of the design but also ensuring that the validation processes reflect the diversity of the user population. This involves conducting extensive testing with diverse groups, addressing potential biases, and making necessary adjustments to improve accuracy and inclusion. By doing so, designers can create systems that are fair, effective, and accessible to everyone, fostering a more equitable technological landscape and minimizing the risk of inadvertently putting certain groups at a disadvantage.

Designing for racial/ethnic diversity requires consideration for Inclusive Design principles as well (discussed in detail in Chapter 3), which involve a deliberate and thoughtful approach to acknowledging and respecting cultural diversity and the unique experiences, values, and perspectives that can be brought to the table with them. For instance, creating inclusive designs could look like providing options that accommodate a variety of language preferences, ensuring that users can interact with products or services in their preferred language. It could also mean having diverse representation in visual elements, such as images and illustrations, to reflect the full spectrum of racial and cultural identities. This representation, however, should be authentic and respectful, avoiding the use of stereotypes or inauthentic cultural references, such as slang that may not resonate with or may be offensive to different groups (I think we've all seen that "Hello you fellow kids" meme, right?).

In the simplest terms, Universal Design aims to provide a single solution or a uniform series of solutions that work for everyone. In contrast, Inclusive Design seeks to address the diverse needs of users by offering multiple solutions tailored to different requirements. While a single solution might be practical for addressing physical differences, the complexity of racial/ethnic diversity and cultural factors associated with it often demands a range of solutions to effectively accommodate all users.

Certainly, the allocation of resources such as time and money are a significant consideration when implementing a single solution or multiple solutions. However, this must be balanced against the potential negative impact of excluding certain populations, as well as the substantial benefits of fostering inclusion. Ensuring that all individuals have access and feel represented is crucial, and the investment in inclusive design can lead to more equitable and positive outcomes for everyone.

In essence, there are aspects we might not yet fully understand, but by adopting, or at least considering, both "Universal Design" and "Inclusive Design" principles, we can strive to create workspaces and other environments that not only address the needs of a diverse user base but also promote respect and inclusion. This intentional approach helps ensure that every user feels valued and properly represented, leading to a more equitable and engaging experience for all.

Other Design Considerations

Whether utilizing Universal Design or Inclusive Design, there are other considerations for designing with racial and cultural diversity that should be considered, specifically within manufacturing and/or office settings. For starters, when it comes to personal protective equipment (PPE) such as uniforms, wearables, exoskeletons, respirators we have discussed how ensuring elements such as adequate sizing are a critical procurement process consideration for gender diversity. Given that historically speaking PPE had been developed from measurement data derived from U.S.'s young, white male military recruits from the 1950s to the 1970s (Robertson et al., 2020), the same applies for racial and cultural diversity needs. These needs, of course, refer to ensuring adequate fit, as due to variations in body proportions across race/ethnicities and/or facial differences (Dempsey, 2023). It is essential to ensure that the styles of PPE and uniforms reflect physical and cultural differences, including the necessity to accommodate beards, head coverings, and various cultural hairstyles such as braids and locs. While safety remains the paramount factor in selecting PPE and designing processes, it is equally important to acknowledge the significance of cultural identity and personal expression. For many individuals, their hairstyles, head coverings, and grooming choices are not just personal preferences; they are vital aspects of their cultural identity and heritage. Providing options that respect and celebrate these differences can foster a more inclusive work environment by way of enhancing employee feelings of acceptance while also demonstrating the organization's commitment to valuing diversity.

In addition to acknowledging and integrating racial and cultural needs in the procurement process and other areas of product and equipment design, it is vital to engage employees in every aspect of the ergonomics and safety process. This includes fostering an environment where open feedback is encouraged, allowing employees to actively participate by identifying specific challenges or requirements within the workforce that must be addressed in relation to design and/or selection of PPE to empower them to embrace their identities without compromising safety.

In the end, we as ergonomics practitioners need to challenge ourselves to think more broadly about the user experience, including potential impacts shaped by a user's racial, cultural, or ethnic identity. Our goal should be to create designs that are not only functional but also equitable. The most effective way to achieve this is through a participatory approach—actively engaging and learning from individuals across a wide range of diverse groups throughout the entire design process. Leveraging insights from an employee-driven ergonomics team can be particularly valuable in this regard, as it offers a direct connection to individuals that interact with the work environment every day, and who have a variety of perspectives and experiences. By doing so, we can ensure that our designs meet diverse needs and contribute to a more inclusive and fairer environment for all.

DESIGNING FOR SOCIOECONOMIC DIVERSITY

Diversity is multifaceted. It is that rich complexity that makes us unique and what makes designing *for* the worker so challenging. There are many characteristics and attributes that can influence and shape an individual's perspective, and consequently how they may react, perceive, engage with a product or process design. Socioeconomic diversity is probably the most challenging diversity characteristic that many ergonomics practitioners struggle to see how they specifically can make an impact. However, while complex, studies have shown (d'Errico et al., 2007; Kim et al., 2014) that there is a negative correlation between socioeconomic factors and musculoskeletal disorders (MSDs), making our role in addressing the disparities paramount.

Socioeconomic diversity is a reflection of the range of differences within groups as related to income, education, occupations, and social class. It is these differences, as with other diversity attributes, that can significantly shape their perspectives and also impact the opportunities and resources available to them, thus leading to disparities in life outcomes. These factors present a variety of both physical and psychological risks and stressors to the individual. As ergonomics practitioners, our role in addressing and mitigating socioeconomic diversity challenges lies in our research, practices and advocacy—in how we aid in improving and mitigating the physical and psychological risk exposure within and around the workplace.

But what are those socioeconomic challenges we may face in the workplace?

Socioeconomic Challenges in the Workplace

For starters, it has been noted that an individual's socioeconomic status (which can be in relation to financial health, education, and perceptions of social class) can significantly influence employment opportunities, including working conditions and individuals within low socioeconomic status more likely to work in environments with hazardous conditions, thus increasing their risk exposure (APA, n.d.a.; APA, n.d.b.). Additionally, studies have shown that risk factors for MSDs, as well as cardiovascular disease, hypertension, diabetes are prevalent in these jobs, such as increased blood pressure with job strain (Landsbergis et al., 2003), fatigue and sleep deprivation (Barger et al., 2005; Cochrane, 2001), smoking (Barbeau et al., 2004; Sorensen et al., 2004), and burnout (Sheiner et al., 2002).

In addition to more challenging working conditions for individuals in the lower socioeconomic statuses, compounding stress exposure is the increased risk of limited autonomy and general control over work pace and schedules. In a 2017 Safety and Health Magazine article, it was reported, as part of the 2015 American Working Conditions Survey where over 3000 workers aged 25 to 71 responded, that a quarter of workers surveyed said they do not have enough time to complete job tasks, more than one-third said they have no control over work schedules, and 66% said they frequently work at a fast pace or under tight deadlines (Safety and Health Magazine, 2017). It is this combination of physical and psychosocial risks that increases the number of MSDs seen with individuals in lower socioeconomic statuses.

Armed with this reality, we as ergonomics practitioners can equip ourselves with mitigation plans to address the socioeconomic work disparities. This would include:

- *Access to Ergo Resources*

It is not a reach to see how access to ergonomics resources, particularly furniture and equipment such as adjustable workstations or sophisticated automation could be limited in lower paying jobs. And this is unfortunate considering how the risk exposure would dictate more of a need for those resources. As we assess and prioritize work within our workplaces, we need to ensure that we are taking a holistic view at the application of our resources. Are we ensuring that budgeting is adequate, especially for workspaces that traditionally require high physical and mental demand due to older equipment? In these cases, a targeted and detailed phase-in plan for improvements over multiple years may be warranted and made a priority. To garner support, a thorough cost-benefit analysis can demonstrate the long-term benefits of investing in improvements, even for older equipment and facilities, highlighting potential cost avoidance from reduced injuries, cost savings from improved operational efficiency, decreased absenteeism/presenteeism and improved employee morale (See Part III to learn more about other metrics that could be used within a cost-benefit analysis).

- *Accessible Ergo Solutions*

As we develop our ergonomics plans to address those priority areas, we also need to be mindful of costs for solutions. In the case of aging equipment, the cost for complete upgrades or retrofits may require significant costs, making progress a challenge when the cost-benefit analysis is not clear. While effort should continue to be made on longer-term equipment improvement plans, we also need to have a frugal mindset, especially for interim plans.

There is definitely a benefit to focusing on high-tech solutions; there can be improved efficiency, accuracy, and performance as well as a better ability to meet any updated regulatory or compliance standards that require specific technological capabilities or features within machinery or equipment. And then there is the enhanced ability to be able to scale for future demands. Yet, there is still a place for frugal yet effective solutions, especially in the interim as companies develop plans and obtain resources to implement high-tech solutions. Better yet, there is an advantage to having a balanced approach that incorporates both high-tech and frugal solutions where appropriate. And cost should not hinder or prevent people from good ergonomics—meaning, we should emphasize the importance of making ergonomics solutions accessible and affordable for everyone, regardless of their financial situation.

Take for example the modern wheelchair. If you imagine the wheelchair of yesteryear, you will more than likely envision a wicker chair with wheels. Not much has changed over the years, unfortunately. While there have been innovations in wheelchair design, many efforts focus on advanced technologies—like power assist features, robotic components, and smart functionalities—that significantly increase costs compared to traditional manual or basic motorized models. And for individuals with limited financial resources, they may struggle to afford these higher-priced options thus pricing them out of valuable innovation. All this to say that there is still ample opportunity for innovation that also takes into account the needs of those from low socioeconomic backgrounds. By prioritizing both high-tech and affordable solutions, we can ensure that progress benefits a wider range of users.

Ergonomics programs are strong proponents and facilitators of what is known as "frugal innovation." Frugal innovation is a concept that emphasizes simplicity, affordability, and accessibility in problem-solving. This approach involves finding innovative solutions within limited resources and constraints, often necessitating the redesign of products, processes, machines, and tools with minimal resources (Khan, 2016)—such as redesigning products (processes, machines, tools, etc..) out of necessity and with minimal resources, which is very often the case. It is this scarcity of resources that drives creative problem-solving. Additionally, frugal innovation typically comes from a bottom-up employee-driven approach, meaning it starts at the bottom in terms of its focus and then expands to benefit all users. In short, this approach to innovation further challenges traditional norms, celebrates resourcefulness, and aims to benefit the many with functional solutions.

One only has to look at all the great solutions that are showcased yearly at the annual Applied Ergonomics Conference Ergo Cup to see how creative, cost-effective solutions can make a huge impact on the operational efficiency, accessibility and safety of employees across the globe. The competition itself embodies the ongoing challenge of achieving cost-effective ergonomics solutions, motivating participants to harness their creativity in developing impactful innovations. Participants in the Ergo Cup leverage their ingenuity to tackle this challenge head-on, devising practical solutions that not only address safety and inefficiency concerns but also prioritize cost-effectiveness. These solutions serve as inspiring examples of how thoughtful design and resourcefulness can revolutionize workplace environments, benefiting organizations by improving productivity and reducing the risk of injuries. By highlighting these achievements, the Ergo Cup highlights the importance of accessible solutions and how strategic thinking and innovation can drive positive change in workplaces across diverse workgroups and industries. This emphasis on cost-effectiveness not only encourages practical problem-solving but also encourages a culture of continuous improvement in ergonomics practices globally.

As ergonomics practitioners, it is imperative that we adopt a similar mindset in all our endeavors, striving to be conscientious stewards of costs within the industries we serve. Our objective extends beyond mere financial prudence; it is about making ergonomics universally accessible. We should aim to integrate cost-effective strategies into our mitigation efforts and make those solutions accessible to all. This will help to ensure that every individual, regardless of economic resources or socioeconomic status, will be able to benefit from optimized work environments that enhance safety, well-being and productivity. Advocating for accessible ergonomics speaks to our commitment in creating sustainable, inclusive workplaces where everyone can thrive. Although this increases our challenge of developing cost-effective solutions that meet the myriads of needs of a diverse workforce, it is a challenge that will drive us to continuously innovate.

One way to embed that into our everyday activities for our ergonomics processes is to establish prioritization criteria that specifically consider cost-effectiveness through ease of implementation. A valuable tool for this approach is a Priority Matrix, which facilitates quick comparison of potential solutions (such as those developed via a solution development event). This matrix utilizes one axis to evaluate the effectiveness of reducing risks associated with ergonomics-related injuries, while the other axis assesses the ease of implementation (Figures 11.1 and 11.2).

The criteria used to gauge ease of implementation may vary based on specific requirements, but common factors include cost, resource availability, time constraints, and the anticipated level of employee buy-in or acceptance. By systematically evaluating solutions against these criteria, ergonomics practitioners can pinpoint interventions that not only promise significant risk reduction but also align with practical considerations of affordability and feasibility. For example, a proposed solution might score high in effectiveness for risk reduction but could prove challenging in terms of cost or resources required for implementation. Conversely, another solution might offer moderate effectiveness but excel in terms of being frugal and straightforward to implement, potentially making it a more suitable choice—or a suitable short term solution— depending on the context and available resources.

By employing this structured approach, ergonomics practitioners can optimize decision-making processes, ensuring that selected interventions not only deliver tangible benefits in terms of health and safety but also be cost-effective. This methodology fosters a balanced approach where the goals of enhancing workplace ergonomics and optimizing accessibility for all employees can be achieved.

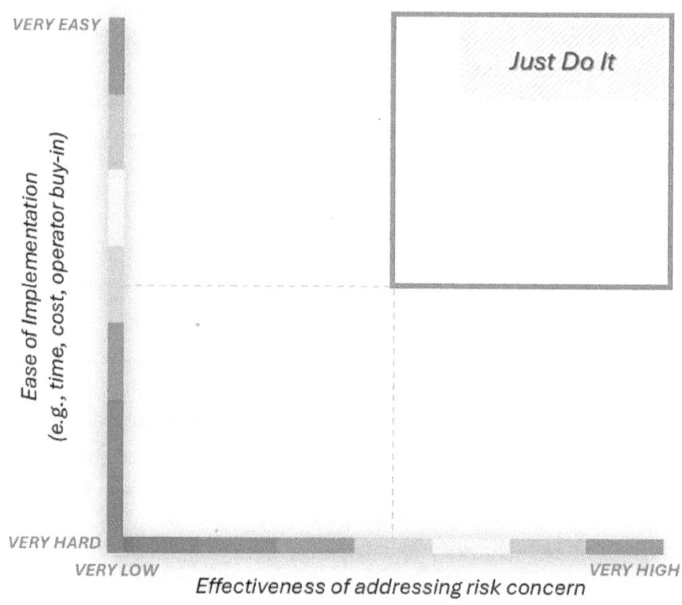

FIGURE 11.1 Priority Matrix. (source: author)

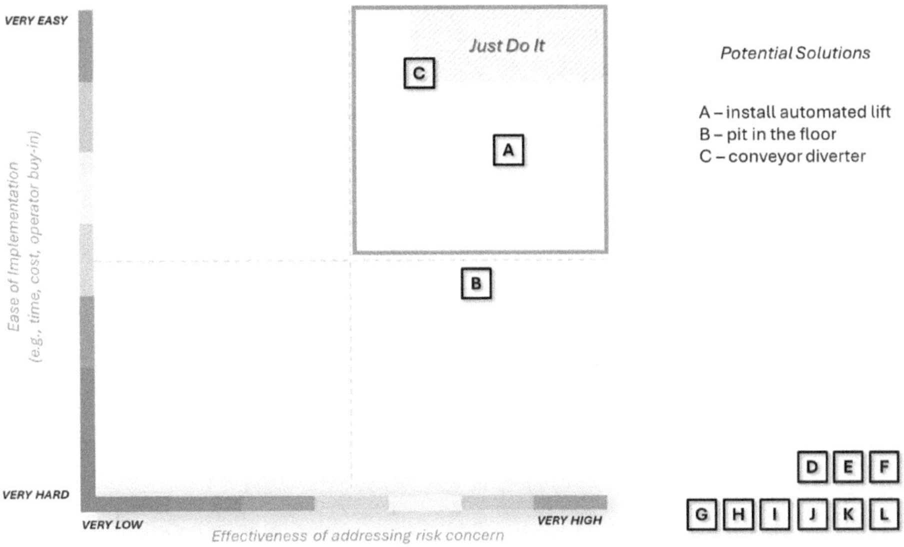

FIGURE 11.2 Example of partially completed Priority Matrix. (source: author)

- *Accessible Ergonomics Training*

In terms of training, ergonomics practitioners should not only ensure that ergonomics training is accessible—meaning it is shared and disseminated in a way that is accessible and understandable to people with varying levels of education and economic resources—but that the content emphasizes the challenges, as associated with ergonomics, as well as potential mitigations.

Suggestions on how to make training more accessible:

- Develop and tailor training for specific groups, such as training specific for (see Table 11.1):
  - Operators/Employees.
  - Management.
  - Engineering (Process, Machine/Equipment, Design).
  - Ergo Team.
- Incorporate visual aids such as infographics, diagrams, and videos to enhance understanding, particularly for visual learners as they clarify complex ergonomics concepts and demonstrate proper techniques clearly and concisely. Ensure that all visual materials used in training are designed with consideration for colorblind individuals. Utilize color schemes that are distinguishable for those with color vision deficiencies and incorporate alternative methods such as patterns or labels to convey information effectively to all participants.
- Offer training via various mediums, such as interactive in-person sessions with instructors for hands-on demonstrations, mobile-friendly online

TABLE 11.1
Suggested ergonomics training content (source: author)

	Operators	Management	Engineers	Ergo Team
Basic Ergo Awareness	✓	✓	✓	✓
Definition of Ergo				
MSD and risk identification				
Ergo principles for solution development				
Ergo KPIs (how measured)				
How to engage with the ergo process				
Cost of MSDs		✓	Optional	✓
Benefits of Ergo		✓	Optional	✓
Regulations and standards (as applicable)		✓	✓	✓
Prevention through design			✓	✓
HMI			✓	✓
Risk assessment (tool training)			Optional	✓
Emerging technology			✓	✓

modules for flexible accessibility, web-based platforms for self-paced learning, and peer-to-peer sessions for collaborative knowledge sharing.

- Consider including gamification in the training such as quizzes, challenges, and rewards motivate participants to actively participate, apply ergonomics principles, and track their progress effectively.
- Offer training materials in multiple languages to accommodate diverse linguistic backgrounds within the workforce; this is crucial for organizations with global operations or multicultural teams as it ensures equitable access to essential ergonomics knowledge and practices (this may be required depending on the location of your facilities—e.g., if in Quebec Canada, training material is required to be presented in French)
- Provide information on the basics of ergonomics, including the definition of word (and any derivatives), risk factors for MSDs, ergonomics principles and how to apply them. Encourage proactive self-identification and mitigation of risks.
- Include information on how psychosocial risks, such as job stress, job satisfaction, and perceived control over work tasks, can influence how individuals cope with risks and injuries. Share steps they can take to utilize resources such as the company's Employee Assistance Program for

counseling and support, how to practice stress management techniques like mindfulness, regular exercise, and hobbies, and how to establish boundaries between work and personal life to prevent burnout and enhance well-being for a healthier life-work balance.
- Emphasize the importance of early reporting of injuries, including persistent pain and discomfort which could be a precursor symptom of an MSD. Provide information on how to spot signs and symptoms of MSD and the specific steps on how to report injuries and what to expect of the process, including what is expected of them.

Ergonomics, Socioeconomic Status, and The Opioid Epidemic

Another significant issue that needs to be on the minds of all ergonomics practitioners is the opportunity our efforts can have in addressing the opioid epidemic that disproportionately impacts workers with lower socioeconomic status. I've discussed how individuals in the lower socioeconomic statuses are disproportionally working in challenging conditions, including physically demanding jobs. It is this disproportion that leads to an increase in connection between this disparity and the opioid epidemic.

PRO TIP

One source of inequity for frontline workers is that they may not always have equitable access to professional or managerial level training, thus impacting their access to career development opportunities (McKinsey & Company, 2022).

Consider working with your local training department to package ergonomics training as official development training. Highlight not only practical assessment skills but also a range of technical and soft skills as well as leadership competencies. These could include user-centric design, innovation, problem-solving, lean manufacturing, change management, empathetic listening and effective communication. This approach not only aids personal growth and career development for those who may not have access otherwise (due to operational constraints) but also addresses potential inequities in access to professional training for the organization.

The opioid issue (non-medical use and overuse of medical use) is global. According to the United Nations Office on Drugs and Crime (UNODC) 2019 Global Smart Update—"Understanding the Global Opioid Crisis" report (2019), the use of opioids, particularly potent substances like fentanyl and its analogues, poses a serious public health concern, especially in North America, but also in parts of Asia, Europe, and Oceania. The UNODC goes on to state that in recent years, there has been a significant increase in opioid-related overdose deaths and hospitalizations, particularly in North America and, to a lesser extent, Europe. And that a surge of new substances with opioid effects has been reported in 29 countries, further exacerbating the crisis.

Specially within the U.S., where the concern is more prevalent, opioid overdose deaths are significantly affecting workplaces by compromising occupational health and safety, leading to increased workers' compensation and health insurance costs, absenteeism, and lost productivity—according to a public health report from the American Public Health Association (APHA) on the public health approach to protecting workers from opioid use disorder and overdose related to occupational exposure, injury, and stress (APHA, 2021). And beyond the workplace, the opioid crisis is resulting in job loss, family disruptions, and various individual health issues. Pain from occupational injuries and stress contributes to opioid use disorder. The APHA stresses the need to include policy reforms to transform workplaces into environments that promote prevention rather than misuse, such as including policy interventions for preventing workplace injuries and emotional distress, enhancing worker education and training, and replacing punitive substance use programs with supportive measures. Additionally, improving access to alternative pain treatments and preventing opioid misuse within workers' compensation systems are crucial for addressing these challenges. The latter is particularly evident due to socioeconomic disparities, where structural and economic factors—such as insufficient pain management, criminal justice policies, and limited access to addiction treatment and harm reduction services—are identified as components of a socioecological model of the opioid crisis (Saloner et al., 2018). The APHA's report denotes that opioid mortality rates and declines in life expectancy are notably higher among individuals with lower education levels, those in occupations with median incomes under $50,000, and blue-collar workers. Considering the intersectionality of socioeconomic status and racial diversity in relation to workplace risks, the APHA notes that the higher rates of occupational injuries, illnesses, and fatalities among Black and Hispanic workers can be partly attributed to their disproportionate employment in high-risk industries.

In an article published in *The Synergist* (Dale et al., 2019) that examines this connection and the efforts the ergonomics profession can take to address it, the authors share how various studies highlight the vulnerability of workers in construction, mining, and other high-risk sectors to opioid prescriptions and associated deaths, with a significant percentage of opioid-related deaths occurring in industries with high injury rates. The lack of paid sick leave and job security further exacerbates the situation, pushing injured workers back to work while still in pain. They posit that ergonomics efforts could be treated as a primary prevention step through the reduction and severity of workplace injuries. This is because studies have shown a strong correlation between work-related pain treatment and opioid misuse, addiction, and overdose fatalities. For example, 2.1 million people in the U.S. were reported to have an opioid use disorder in 2018, with a primary reason for misuse being the relief of physical pain as reported by the Substance Abuse and Mental Health Services Administration (2019).

Considering the link between workplace injuries, particularly those requiring ongoing pain management, as seen with MSDs, effective ergonomics programs can play a crucial role in preventing or lessening the severity of injuries. These programs could also decrease the reliance on opioids for pain management, ultimately reducing costs related to workers' compensation and enhancing overall worker health. Additionally, it is important to recognize that, given the multi-level intersectionality, ergonomics programs can serve as a valuable strategy for addressing socioeconomic and racial disparities, helping to ensure a more equitable work environment.

Advocacy

Advocacy continues to play a pivotal role in addressing and mitigating socioeconomic disparities. Through advocacy efforts, attention is drawn to the inequalities that exist in access to resources, opportunities, and benefits concerning ergonomics and its progress as well as the impact of socioeconomic factors on workplace health, safety, and productivity. By increasing awareness among stakeholders, we provide valuable context regarding the significance of addressing these disparities. Additionally, it highlights the importance of considering diverse socioeconomic backgrounds in the development and implementation of ergonomics strategies. This inclusive approach not only addresses immediate disparities but also fosters a culture of equity and support within organizations. It encourages proactive measures to ensure that all employees, regardless of their socioeconomic status, benefit from ergonomics efforts that can enhance their overall work experience and quality of life.

The APA has priorities associated with advocacy efforts in public policies that mitigate the effects of scarcity and poverty (APA, n.d.b). Those efforts include, as related to the workplace, a focus on securing funding that promotes research that develops effective solutions that meet the unique needs of low-income people, increasing the earnings of low wage earners, and enhancing overall physical and mental well-being. Our role as ergonomics practitioners, and our profession as a whole, could help serve as a crucial bridge between awareness and action, especially for advocacy within the workplace. We can do this by driving meaningful change toward a more equitable and sustainable workplace environment and emphasizing the importance of collaborative efforts in addressing socioeconomic disparities, promoting fairness, inclusion, and well-being for all individuals in the workforce. Practically speaking, we can do this by providing a platform, through our participatory approach, for workers to voice their concerns, particularly regarding the work environment. Through our ergonomics processes, we can ensure equitable access to ergonomics resources and support, including addressing potential physical and psychosocial risks associated with socioeconomic disparities and ensuring accessibility to solutions. Lastly, our ergonomics process can be used as a catalyst to foster collaboration and innovation for all workers, regardless of socioeconomic status.

DESIGNING FOR LGBTQ+

As organizations strive to foster environments where all employees feel valued and comfortable, recognizing the unique needs of LGBTQ+ individuals is important in work and workplace design. By addressing design challenges, such as inclusive communication and gender-neutral spaces and facilities, companies can ensure that their workspaces accommodate diverse identities and expressions. This thoughtful approach not only enhances the well-being and productivity of LGBTQ+ employees but also reflects a commitment to equality and respect for all members of an organization's workforce.

Inclusive Signage and Communications

Effective communication is essential in any workplace. At its core, it serves to share information that prompts action or influences behavior. More broadly, it is vital for ensuring that work is carried out as designed, based on thorough work studies for efficiency and safety, facilitating effective collaboration among team members, and managing expectations. The effectiveness of communication—whether visual, verbal, or otherwise—depends largely on how well it resonates with the intended audience, which is influenced by the level of empathy and respect it conveys.

This principle applies to everyone receiving, hearing, or interpreting the message, including employees who are members of the LGBTQ+ community. As discussed, psychological safety can be significantly impacted by how work and the work environment are perceived. If work is seen as exclusionary or disrespectful of an individual's experiences, it can lead to feelings of psychological unsafety, with negative consequences as previously mentioned. This includes workplace communication.

But what should be considered when developing communications for the workplace as related to the LGBTQ+ community? A number of resources that provide guidance on how to develop inclusive communication, specifically for the LGBTQ+ community, are available, such as the recommendations from the U.S.'s National Institutes for Health (NIH, n.d.). These include suggestions regarding:

- **Terminology**: The key here is to use respectful and accurate wording and terminology when referring to LGBT+ individuals and communities. Ensure that any terminology used in training materials, policies, or everyday conversations reflects current and respectful usage.
- **Inclusive Language and Symbols**: Embrace gender-neutral language, avoiding assumptions about gender identity or sexual orientation. For instance, use terms like "partner" or "spouse" rather than specifying "husband" or "wife." In professional settings, opt for "everyone" or "team" instead of "ladies and gentlemen" to make the statement more inclusive. For symbols and icons, remember that symbols that are widely recognized

and accepted in one culture might not hold the same significance or be as well received in another.

- **Representation**: Strive for diverse representation in all workplace communications, including a reflection of a broad spectrum of LGBTQ+ identities and experiences. This can be done through the use of inclusive imagery or sharing stories and achievements from LGBTQ+ employees to highlight their contributions and experiences.

Gender as a Diverse Spectrum

Beyond the practicality of designing for gender diversity, the topic itself comes with challenges and barriers. These can include how social norms, which are unwritten rules and expectations that govern the behavior of a group or society, or in this case, the work environment, perpetuate stereotypes in the workspace. These stereotypes can make even discussing changes difficult and can be met with resistance within (and outside) of the organization. Additionally, a political landscape that can lend itself to backlash and outright resistance consequently leading to reluctance on the part of those asked to institute changes for equity.

As society progresses, understanding gender as a diverse spectrum rather than a simple binary has grown in significance. That should mean when policies are discussed, debated, and decided upon, those voices and perspectives should be represented in those decision-making spaces to ensure effective consideration and advocacy.

These challenges necessitate ergonomics practitioners to adopt a mindful design approach that honors and includes all gender identities. For the ergonomics practitioner, this could mean advising on design questions such as the anthropometric considerations that would go into developing shared spaces within the workplace (e.g., gender-neutral bathrooms—which, as advised by Neumann Monson Architects (Chan, 2023), could include the use of single-occupancy rooms which offers a safe and private space regardless or non-gendered group bathrooms with floor to ceiling private stalls and communal sinks).

Or it could mean advising of gender-neutral symbols to use in communications such as surveys designed to solicit feedback without unintentionally biasing responses. As such, as should be the case for all general design topics as it will help us to stay current on how best to ensure our communications resonate with a changing workforce it will be important for us to stay abreast of design discussions in this space, which is currently evolving. For example, given the ongoing global debates around gendered bathrooms the design field is weighing in on the common pictograms used to signal gender bathrooms. Recommendations include ideas ranging from blending the pictograms into one symbol to nixing the clothing (the canonical A-Line dress used in the WOMAN icon) altogether to focus on functionality of the bathroom as opposed to separate genders. Additionally, relying on standard human shape symbols called anthropographics used in design as a default can create misleading interpretations of diverse populations, as noted by Dhawka, et al. (2023). In their work, they have pointed out the implications of using such homogeneous design symbols in trying to represent human variability, including potentially obscuring important differences between groups or perpetuating harmful stereotypes. Despite the ongoing debates in and around this area of design, this is a good example of how language, including simple visuals, has the power to include or exclude.

Ultimately, as with all the topics in this book, these design challenges require us to educate ourselves in this area and, more importantly (in my view), to embrace openness. This requires intentionality. We must be intentional in learning about various gender identities and expressions, use gender-neutral language in our communications, and actively avoid making assumptions based on gender stereotypes (such as the simplistic "shrink it and pink it" approach when designing for women).

These challenges also highlight the importance of adopting a collaborative and intersectional approach to designing for gender diversity—including consideration for gender as a spectrum—ensuring that all voices are heard and valued in the design process. Beyond our own personal growth in understanding this topic, this area presents a great opportunity for ergonomics practitioners to contribute. We can add value by incorporating diverse experiences and perspectives, including those of all genders, through user-centered design approaches. Most importantly, we can advocate for inclusive design to C-suite decision-makers, highlighting how inclusion enhances safety, well-being, productivity, and quality—benefits for everyone.

Practically Speaking

It is important to remember that perfection isn't always attainable; we are all prone to making mistakes. A key part of our work is acknowledging when we fall short and taking steps to improve. While it can be challenging and possibly a bit overwhelming to consider all the variables involved, this is a fundamental aspect of being an ergonomics practitioner.

Our goal is to design work environments that accommodate the diverse needs of workers, no matter how complex. We must also recognize and help others understand that every decision we make—whether it involves the language we use or the functionalities we incorporate into processes—can either foster inclusion or lead to exclusion. This awareness points out the value of our work and the impact we can have. To develop truly inclusive designs, it is crucial for us to engage with our diverse work communities and build relationships with partners from the LGBTQ+ community and other diverse groups for which we are designing.

FURTHER READING

For additional reading on designing and mitigating for challenges related to racial, cultural or socioeconomic diversity or considerations designing for the LGBTQ+ community consider the following:

- Harvard Business School article: "How to Promote Racial Equity in the Workplace" (Livingston, 2020).
- Should Data On Gender and Ethnicity Inform Ergonomics Interventions? Lessons from Four Case Studies (Messing & Blanchette-Luong, 2022).

- American Psychology Association: Education and Socioeconomic Status (APA, n.d.c.).
- Centers for Disease Control: Socioeconomic Factors (CDC, n.d.).

REFERENCES

American Psychological Association (APA). (n.d.a). Work stress and health. Retrieved June 9, 2024, from https://www.apa.org/pi/ses/resources/publications/work-stress-health

American Psychological Association (APA). (n.d.b). *Education and socioeconomic status.* American Public Health Association. Retrieved November 13, 2024, from https://www.apa.org/pi/ses/resources/publications/education

American Psychological Association (APA). (n.d.c). *Socioeconomic status and health.* APA Services. Retrieved July 9, 2024, from https://www.apaservices.org/advocacy/issues/socioeconomic-status

American Public Health Association (APHA). (2021, January 13). Protecting workers from opioid use disorder. American Public Health Association. https://www.apha.org/policies-and-advocacy/public-health-policy-statements/policy-database/2021/01/13/protecting-workers-from-opioid-use-disorder

Barger, L. K., Cade, B. E., Ayas, N. T., Cronin, J. W., Rosner, B., Speizer, F. E., & Czeisler, C. A. (2005). Extended work shifts and the risk of motor vehicle crashes among interns. *New England Journal of Medicine*, *352*(2), 125–134. https://doi.org/10.1056/NEJMoa041401

Barbeau, E. M., Krieger, N., & Soobader, M. J. (2004). Working class matters: Socioeconomic disadvantage, race/ethnicity, gender, and smoking in NHIS 2000. *American Journal of Public Health*, *94*(2), 269–278.

Brar, V., Kumar, A., & Wadajkar, V. (2019). Impact of workplace discrimination on employee wellbeing. *VSRD International Journal of Research & Development*, *9*(2), page range. https://www.viirj.org/vol9issue2/40.pdf

Buolamwini, J., & Gebru, T. (2018). Gender shades: Intersectional accuracy disparities in commercial gender classification. In *Proceedings of the 1st Conference on Fairness, Accountability, and Transparency* (pp. 77–91). PMLR. https://proceedings.mlr.press/v81/buolamwini18a/buolamwini18a.pdf

Centers for Disease Control and Prevention. (n.d.). *Socioeconomic factors and health equity.* National Center for Chronic Disease Prevention and Health Promotion. Retrieved November 13, 2024, from https://www.cdc.gov/dhdsp/health_equity/socioeconomic.htm

Chan, C. (2023, June 1). *Gender-neutral restrooms: Types & considerations.* Neumann Monson. https://neumannmonson.com/blog/gender-neutral-restrooms-types-considerations

Cochrane, G. (2001). The effects of sleep deprivation. *FBI Law Enforcement Bulletin*, *70*, 22.

Cordero, E. D., Loredo, J. S., Murray, K. E., & Dimsdale, J. E. (2012). Characterizing fatigue: The effects of ethnicity and acculturation. *Journal of Applied Biobehavioral Research*, *17*(1), 59–78. https://doi.org/10.1111/j.1751-9861.2012.00077.x

Dale, A., Evanoff, B., Macomber, M., O'reilly, M., Rosen, J., & Schneider, S. (2019). *Ergonomics and the opioid crisis.* Synergist. https://synergist.aiha.org/201905-ergonomics-opioid-crisis

d'Errico, A., Punnett, L., Cifuentes, M., Boyer, J., Tessler, J., Gore, R., Scollin, P., Slatin, C., & Promoting Healthy and Safe Employment in Healthcare Research Team. (2007). Hospital injury rates in relation to socioeconomic status and working conditions. *Occupational and Environmental Medicine*, *64*(5), 325–333. https://doi.org/10.1136/oem.2006.027839

Dempsey, P. G. (2023). Human factors/ergonomics contributions to an equitable personal protective equipment protections strategy. *Proceedings of the Human Factors and Ergonomics Society Annual Meeting, 67*(1), 816–820. https://doi.org/10.1177/21695067231192598

Dhawka, P., He, H. A., & Willett, W. (2023). We are the data: Challenges and opportunities for creating demographically diverse anthropographics. In *Proceedings of the 2023 CHI Conference on Human Factors in Computing Systems (CHI '23)*, 807, 1–14. Association for Computing Machinery. https://doi.org/10.1145/3544548.3581086

Edmondson, A. C. (2020, June 22). The role of psychological safety in diversity and inclusion. *Psychology Today.* https://www.psychologytoday.com/us/blog/the-fearless-organization/202006/the-role-of-psychological-safety-in-diversity-and-inclusion

Estafanous, A., & Sedky, K. (2023). Culture, ethnicity, and sleep patterns and disorders. In *The WASP textbook on social psychiatry: Historical, developmental, cultural, and clinical perspectives* (p. 410). Oxford University Press.

Flynn, M., Castellanos, E., & Flores-Andrade, A. (2018). Safety across cultures: Understanding the challenges. *Professional Safety, 63*, 28–32.

Khan, R. (2016). How frugal innovation promotes social sustainability. *Sustainability*, 8(10), 1034. https://doi.org/10.3390/su8101034

Karasz, A., & McKinley, P. S. (2007). Cultural differences in conceptual models of everyday fatigue: A vignette study. *Journal of Health Psychology, 12*(4), 613–626. https://doi.org/10.1177/1359105307078168

Kim, S. K., Kim, H., Lee, K., et al. (2014). The relationship between injury and socioeconomic status in reference to the Fourth Korean National Health and Nutrition Examination Survey. *Annals of Occupational and Environmental Medicine, 26*, 1. https://doi.org/10.1186/2052-4374-26-1

Koenecke, A., Nam, A., Lake, E., Nudell, J., Quartey, M., Mengesha, Z., Toups, C., Rickford, J. R., Jurafsky, D., & Goel, S. (2020). Racial disparities in automated speech recognition. *Proceedings of the National Academy of Sciences of the United States of America, 117*(14), 7684–7689. https://doi.org/10.1073/pnas.1915768117

Landsbergis, P. A., Schnall, P. L., Pickering, T. G., Warren, K., & Schwartz, J. E. (2003). Lower socioeconomic status among men in relation to the association between job strain and blood pressure. *Scandinavian Journal of Work, Environment & Health, 29*(3), 206–215. https://doi.org/10.5271/sjweh.723

LeSavage, H. (2024, September 25). Youthforia foundation scandal: The darkest shade controversy. *Marie Claire.* https://www.marieclaire.com/beauty/youthforia-foundation-darkest-shade-scandal/

Livingston, R. (2020). How to promote racial equity in the workplace. *Harvard Business Review.* https://hbr.org/2020/09/how-to-promote-racial-equity-in-the-workplace

McKinsey & Company. (2022, July 30). Race in the workplace: The frontline experience. https://www.mckinsey.com/featured-insights/diversity-and-inclusion/Race-in-the-workplace-The-frontline-experience#/

Messing, K., Chadoin, M., & Blanchette-Luong, V. (2022). Should data on gender and ethnicity inform ergonomics interventions? Lessons from four case studies. *Ergonomics, 65*(11), 1567–1577. https://doi.org/10.1080/00140139.2022.2074098

Mottaghi, Z., Halvani, G., Jambarsang, S., & Mehrparvar, A. H. (2024). Effect of ergonomic intervention on cognitive function of office workers. *Indian Journal of Occupational and Environmental Medicine, 28*(4), 267–271. https://doi.org/10.4103/ijoem.ijoem_213_22

National Institute of Standards and Technology. (2019). *Cybersecurity framework manufacturing profile: A supplement to the NIST cybersecurity framework (NIST IR 8280).* National Institute of Standards and Technology. https://nvlpubs.nist.gov/nistpubs/ir/2019/NIST.IR.8280.pdf

National Institutes of Health (NIH). (n.d.). *LGBTQI+ terminology*. Equality, Diversity, and Inclusion. Retrieved September 23, 2024, from https://www.edi.nih.gov/our-communities/sexual-and-gender-minority/resources/lgbtiq-terminology

Presser, H. B. (2003). Race-ethnic and gender differences in nonstandard work shifts. *Work and Occupations*, *30*(4), 412–439. https://doi.org/10.1177/0730888403256055

Robertson, L. D., Syron, L., Flynn, M., Teske, T., Hsiao, H., Lu, J., & Lowe, B. D. (2020, December 14). Exoskeletons and health equity: Advancing opportunities for all workers. NIOSH Science Blog. Centers for Disease Control and Prevention. https://blogs.cdc.gov/niosh-science-blog/2020/12/14/exoskeletons-health-equity/

Saloner, B., McGinty, E. E., Beletsky, L., et al. (2018). A public health strategy for the opioid crisis. *Public Health Reports*, *133*(suppl 1), 24S–34S. https://doi.org/10.1177/0033354918813300105

Safety+Health Magazine. (2017, August 21). Work conditions unpleasant, potentially hazardous for more than half of Americans, study finds. *Safety+Health Magazine*. https://www.safetyandhealthmagazine.com/articles/16062-work-conditions-unpleasant-potentially-hazardous-for-more-than-half-of-americans-study

Sephton, S. E., & Kay, D. B. (2024). How do psychosocial and cultural factors influence sleep and circadian health disparities? *Research Directions: Sleep Psychology*, *1*, e1. https://doi.org/10.1017/slp.2023.3

Sheiner, E. K., Sheiner, E., Carel, R., Potashnik, G., & Shoham-Vardi, I. (2002). Potential association between male infertility and occupational psychological stress. *Journal of Occupational and Environmental Medicine*, *44*(12), 1093–1099.

Sorensen, G., Barbeau, E., Hunt, M. K., & Emmons, K. (2004). Reducing social disparities in tobacco use: A social-contextual model for reducing tobacco use among blue-collar workers. *American Journal of Public Health*, *94*(2), 230–239.

Substance Abuse and Mental Health Services Administration (SAMHSA). (2019). *Key substance use and mental health indicators in the United States: Results from the 2018 National Survey on Drug Use and Health* (HHS Publication No. PEP19-5068, NSDUH Series H-54). Center for Behavioral Health Statistics and Quality, Substance Abuse and Mental Health Services Administration. https://www.samhsa.gov/data/

Sue, D. W., & Sue, D. (2012). *Counseling the culturally diverse: Theory and practice* (6th ed.). John Wiley & Sons.

Sudat, S. E. K., Wesson, P., Rhoads, K. F., Brown, S., Aboelata, N., Pressman, A. R., Mani, A., & Azar, K. M. J. (2023). Racial disparities in pulse oximeter device inaccuracy and estimated clinical impact on COVID-19 treatment course. *American Journal of Epidemiology*, *192*(5), 703–713. https://doi.org/10.1093/aje/kwac164

The Associated Press (AP). (2023, September 25). Black plaintiffs file lawsuits over wrongful arrests, jailing due to facial recognition. *NBC News*. https://www.nbcnews.com/news/nbcblk/black-plaintiffs-file-lawsuits-wrongful-arrests-jailing-due-facial-rec-rcna117168

Xu, Y. E., & Chopik, W. J. (2020). Identifying moderators in the link between workplace discrimination and health/well-being. *Frontiers in Psychology*, *11*, 458. https://doi.org/10.3389/fpsyg.2020.00458

12 Ergonomics and DEI in Communications

I'm sure that data is fascinating, but I can't read that chart.

That was my coworker's blunt remark as I shared a series of charts I had meticulously crafted for an upcoming team meeting. I had invested hours analyzing data and ensuring everything was accurately represented using a popular business analytics tool. I felt a swell of pride, having dedicated much of my week to making sure the insights the data would show were compelling enough to spur action from the sites—our end users. But when my coworker said that, it felt like a punch to the gut. It stung my ego to realize that, despite being an "expert" in ergonomics, I was about to present something that wasn't accessible to everyone.

The reason he was struggling to decipher the charts? He is colorblind. My choice to rely heavily on color-coded data visuals rendered my work practically unusable for him.

While the comment hit hard, I was grateful he felt comfortable enough to speak up. It was a learning moment I didn't see coming. I planned to share the charts with our central staff as a preview before the larger rollout, so his feedback gave me the chance to rethink my approach. At that point, I wasn't even aware of features within the tool, such as color-blind-friendly palettes, that could have improved my visuals. Ultimately, the oversight was mine for not considering my audience.

This experience brought me back to the basics of ergonomics: *The importance of understanding the needs of your users.* So, I decided to turn this lesson into a broader learning opportunity for others. Every month, I shared ergonomics tips with the organization in the form of monthly posters (both digital and hard copy versions shared via standard company communication channels), and I chose to focus on color blindness for the next topic. I detailed its prevalence, highlighted why it is essential to consider this condition when communicating, and offered practical guidance for making information accessible (all to be shared later).

The response was heartwarming. Many colleagues reached out, expressing their appreciation for shedding light on a topic they often navigated in silence. One particularly touching message from my colleague Nicolas Brindel stood out:

> Being colorblind myself, I've never seen an official message addressing how to accommodate those of us with this condition. I've always been the one reminding others that about 8% of the male population is colorblind to some degree. The Ergo & Colorblindness communication was incredibly well done, and it finally showed me that this difficulty I live with every day is not ignored. I want to congratulate everyone involved in putting it together... colorblind individuals have learned to create their own strategies and may grow weary of constantly mentioning their challenges....

DOI: 10.1201/9781032658988-15

I cherished Nicolas's feedback for two reasons: First, it validated that people actually read the communications I create! Even reaching one person makes my efforts worthwhile. Second, I hoped that other ergonomics practitioners would recognize how impactful it is to acknowledge employees' needs. In that moment, I had turned a simple misstep into an opportunity to create an advocate for inclusion... and ergonomics! That's the power of listening and learning in our field!

So, how should we account for colorblindness in our ergonomics efforts?

For starters, let's talk about what it is. According to the American Optometric Association, color blindness is a color vision deficiency condition in which a person has difficulty identifying a particular shade of color, causing confusion in visual interfaces. This is due to when the cones (photoreceptors) within the retina do not have a sufficient number of light-sensitive pigments, making it difficult for a person to register different colors.

The most common form of color deficiency is red-green, meaning those individuals have a hard time differentiating between the two colors. Blue-yellow color deficiency is rarer and more severe, as they typically have red-green color deficiency. In this case, individuals see neutral or gray areas in place of color. Lastly, those that are totally color deficient are referred to as having "achromatopsia," which is where these individuals can only see things in black and white or in shades of gray. All cases of color blindness can range from mild to severe and can affect both eyes (normally the case if the condition is inherited). However, injuries or diseases (such as diabetes or glaucoma) that affect the optic nerve or retina can also lead to color deficiency, making it difficult to recognize and distinguish colors.

In terms of prevalence, as per the National Eye Institute, approximately 8% of men and 0.5% of women have a color vision deficiency (color blind). Additionally, many individuals with color blindness may not realize that they see colors differently from others. Therefore, chances are this is a condition to account for within the workforce you support.

You can see how a color deficiency condition can lead to miscommunication and possible safety issues if critical communications are not designed effectively to account for it. It is not just likely, but a foregone conclusion that ergonomics practitioners will have to consider the needs of individuals with color deficiencies. However, this is just one of many potential needs that may need to be taken into account when developing and designing communications and communication plans for a diverse workforce.

ACCOUNTING FOR DIVERSITY IN COMMUNICATIONS

The effectiveness and sustainability of an ergonomics process significantly depends on how effective we are at communicating to and with our audiences—be it frontline or office workers, middle and senior management, or end-users of our products and processes. And each of those audiences has unique needs and perspectives that we need to consider. Whether we are communicating general informative details, data in the form of charts and graphs to drive action/behavior, specific instructions, or warnings of hazards, reaching broad audiences that include a diverse workforce requires intentionality in how we craft and design those communications.

Understanding how information flows and is processed by an audience can guide us in shaping our communications for maximum effectiveness. Effective communication requires that the message is noticed, engaged with (perceived), comprehended, and motivating enough to persuade the desired behavior. For instance, consider a common ergonomics tip for handling objects:

Don't twist; pivot your feet instead.

This advice can be communicated in various ways—visually through slides, verbally by an instructor, with illustrations showing the right and wrong techniques, via hands-on demonstrations in a body mechanics course, posted on a billboard in the break room, etc. Wearable devices could also provide tactile feedback when the tip is followed or ignored.

With so many avenues to convey messages, how can we ensure they stand out and aren't lost among the onslaught of other communications workers encounter daily? How can we guarantee that the message is understood, especially in a workforce with diverse primary languages? And how can we make sure it resonates in a culturally diverse environment to the point it motivates the desired behavior (or behavior change)? As the audience evolves, so too should our communication strategies.

While many models exist in the literature on how humans process information—especially regarding safety and risk communication, such as the Protective Action Decision Model (Lindell & Perry, 2012), Risk Information Seeking and Processing Model (Yang et al., 2014), or the Heuristic-Systematic Model (Todorov & Henderson, 2002), the C-HIP model framework (Wogalter, 2006a) is particularly relevant for safety communications. This model highlights that at any stage of the information flow from the source, there can be breakdowns that prevent the message from progressing and fail to elicit the intended behavior. From a diverse workforce perspective, the diverse needs (as shaped by varying experiences and expectations) can influence where the breakdown in the communication flow can happen.

Our challenge in ergonomics-related communications is to reduce the "noise" that can obstruct how messages are received and understood, ultimately affecting the likelihood of desired actions being taken. This is particularly difficult when our audiences have diverse physical, cognitive, and psychosocial needs that must be addressed to effectively reach them.

We need to consider various factors, including cultural attitudes, language preferences and proficiency, physical and cognitive disabilities, work experience levels for task-specific information, and psychosocial elements influenced by cultural, racial, and socioeconomic factors—such as trust, bias (both conscious and unconscious), engagement, and representation. All of these can have an impact on how effectively a form of communication is perceived, understood, and how it resonates with individuals.

To address these challenges, we must be intentional in our design choices regarding:

- The channels and formats used to deliver our communications.
- The language used, taking into account varying proficiencies and learning styles.
- The cultural preferences for how communication is presented; e.g., a preference for general guidelines with freedom to adjust or a preference for specific, detailed instructions or prescriptions to follow.
- The cultural appropriateness and representation of the message and imagery.
- The motivational aspects of the message, acknowledging that attitudes and beliefs will differ.

PRACTICALLY SPEAKING

Designing for Color Deficiencies

Returning to the earlier discussion about individuals with color blindness, there are specific recommendations available regarding how we should design our communications to account for color vision deficiency. For starters, it is essential not to rely solely on color to convey important information within our communications. For example, instead of using color primarily to differentiate between categories in the chart I presented, I could have also incorporated patterns or textures. Imagine a bar graph where one set of bars is a solid color, while another set is dotted or striped. This way, even someone who struggles with color differentiation can better distinguish between the two.

Additionally, consider using other indicators, especially in digital contexts. For instance, if you're displaying critical information on a screen, you could add flashing or blinking elements to grab the audience's attention. Think of a notification system in an app where a message not only changes color but also flashes to alert users to an important action or update. This dual approach makes it much more likely that *everyone* will notice the essential information.

Other considerations to enhance accessibility for those with color vision deficiencies:

- **High Contrast**: Always use high-contrast combinations, like dark text on a light background. This simple combination can significantly improve readability for everyone, not just those with forms of color blindness (Note: tools such as https://accessibleweb.com/color-contrast-checker/ are available).
- **Color-Action Associations**: When using color to signify actions, like changing text to green to indicate completion in an app, it is vital to add another cue. For example, you could pair the green text with a checkmark icon or even a flashing light. This way, even if someone doesn't notice the color change, they can still understand what it means. However, cultural context is important to remember as colors can have varied meanings across cultures (Light Adviser, 2024). For instance, while white may symbolize purity in some cultures, it can signify mourning in others. This cultural variability highlights the necessity of understanding the context in which

colors are used as relevant to the workforce, ensuring that the intended message is communicated more effectively across diverse audiences.

- **Alternative Symbols**: Instead of relying solely on color-coded signals, consider incorporating symbols or shapes to communicate information. For instance, in a safety context, a stop sign could be red and take on an octagonal shape to reinforce its message. However, just as with color "meanings" in various cultures, it is essential to recognize that different countries and cultures may interpret symbols in various ways. Make it a practice to provide a brief description or alt text for images and symbols. This approach not only makes visual aids more inclusive for individuals with visual impairments (as the description can be read by assistive devices) but also helps to ensure that the intended meaning of the image or symbol is clearly communicated.

Other DEI Considerations in Communications—Color

Beyond designing for individuals with colorblindness, there are other considerations to help ensure your communications are more accessible and inclusive. As ergonomics practitioners, we are often responsible for considering every element of our designs that might affect the user—whether physiological, psychological, or emotional. For example, the use of color is important when combined with the previous recommendations to accommodate those with color deficiencies.

Color can play a significant role in shaping the usability of the processes, products, and systems we design, impacting effectiveness, efficiency, and overall satisfaction as detailed by ISO 9241-11. It is not just an aesthetic choice. Color is a fundamental element that can significantly enhance or hinder the user experience.

As noted before, consider how color can be effectively used to convey or code information users will need to discern messages, detect differences, and act appropriately. In user interfaces, specific colors are often employed to communicate messages or prompt actions. We discussed how green can be used to signify that an action has been completed due to the common association of "go" or "safe to proceed" or "successfully completed" (Mulholland, 2020). In contrast, the color red often alerts users to errors or warnings due to its association with "stop" or "do not proceed" (Mulholland, 2020) or "error" (Sharma, 2023).

Given the critical role color plays in designing products, workspaces, and communication, ergonomics practitioners must consider several important aspects of color ergonomics to avoid inadvertently excluding users.

For example, in terms of perceptibility and positioning, warm colors like red, orange, and yellow are more noticeable than cooler tones like blue and green, which is why a bright yellow warning sign will likely catch someone's eye much faster than a sign in a cooler shade (Cios233, 2022). Plus, when it comes to peripheral vision, which is more attuned to black and white, color can be harder to see at the edges of our vision—this is especially important for signage that needs to be viewed from different angles.

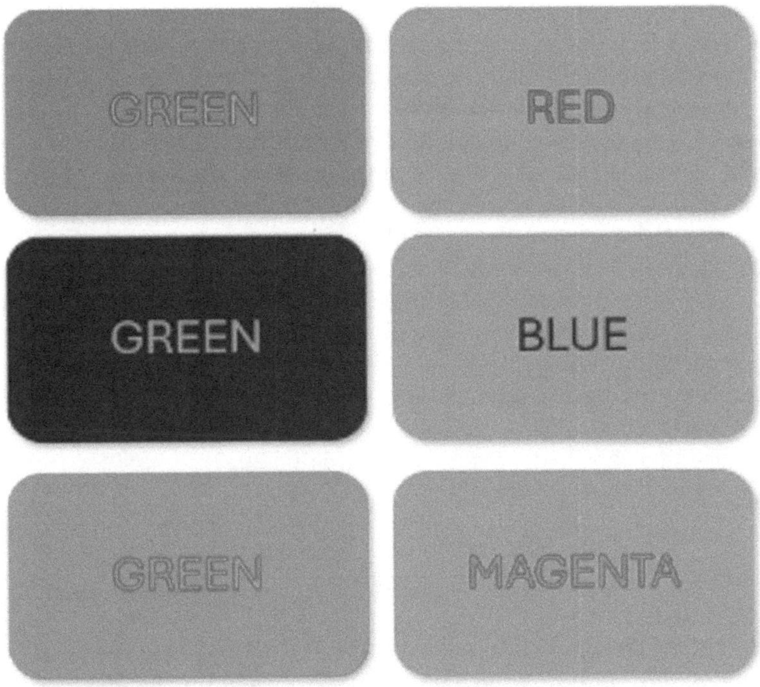

FIGURE 12.1 Image of six color boxes with various colored fonts (source: author); Example of color contrasts with fonts. Note: Color-based contrasts shown in this figure may not be visible if viewing a grayscale print. For an accessible, full-color version, please refer to the digital edition.

Then there's the issue of color combinations; while bright colors can grab attention, certain pairings might create a distracting effect that feels like they're pulsing or shaking (Penn State University, n.d.; see Figure 12.1)—which can distract or irritate users over time. Lastly, as people age (as discussed in Chapter 10), their ability to perceive colors can diminish, making it challenging to comprehend the message when color is the primary means of conveying information (Ishihara et al., 2001). Incorporating other methods, such as shapes or text labels, can enhance clarity and accessibility for older adults.

Other DEI Considerations in Communications—Font, Lettering, Content, and Structure

While color blindness is a physical impairment, conditions such as dyslexia and dyscalculia that can make reading and comprehending visual communications a challenge are neurodivergent conditions in which information is processed differently in the brain (as discussed in Chapter 9). Regardless, there are things to consider within your communications to help avoid making it more difficult for those with these conditions.

Dyslexia is a specific learning disorder/difference that can be described as a continuum of difficulties in learning to read, write, and/or spell. While the condition can present itself differently in each person, common signs (for adults) could include confusing visually similar words, erratic spelling, or difficulty skimming text (British Dyslexia Association, n.d.). It is often associated with difficulties such as auditory and visual processing of language-based information, phonological awareness, and reading fluency. When it comes to designing communications, the British Dyslexia Association (www.bdadyslexia.org.uk) offers the following guidance (British Dyslexia Association, n.d.):

Dyscalculia, like dyslexia, is a specific learning disorder/difference that impacts an individual's ability to understand and work with numbers, significantly hindering their mathematical skills, ability to tell and manage time, handling change, or remembering number facts (National Center for Learning Disabilities (NCLD), n.d.). Here are a few recommendations to consider when sharing information to help those with this condition cope better (McFayden & Malik, 2022):

TABLE 12.1
Communication Design Guidance from the British Dyslexia Association For More Information on the Dyslexia Style Guide, see www.bdadyslexia.org.uk

Topic	Recommendations
Readable fonts	• Use sans serif fonts (e.g., Arial, Tahoma) as letters can appear less crowded Alternatives: Verdana, Century Gothic, Trebuchet, Calibri, Open Sans. • Consider making editable options for PDF documents so that individuals with dyslexia can make changes as needed for better readability. • Font size: 12–14 points (larger for dyslexic readers). • Larger inter-letter spacing (~35% of average letter width); inter-word spacing: At least 3.5 times inter-letter spacing. • Line spacing: 1.5/150% proportional to inter-word spacing. • Use bold for emphasis; avoid underlining and italics to reduce crowding. • Use lowercase letters; avoid all capital letters for continuous text.
Layout	• Left align text; do not justify. • Avoid multiple columns (like in newspapers). • Write short, simple sentences (60 to 70 characters optimal). • Use regular section headings and include a table of contents in long documents.
Writing style	• Use active voice rather than passive. • Be concise. • Write in simple, clear language. • Use images to support text (e.g., flow charts for procedures, pictograms for information). • Use bullet points and numbering instead of continuous prose. • Avoid abbreviations; always provide the expanded form when first used. • Provide a glossary for abbreviations and jargon.

- Round numbers to the nearest whole number, using decimals (.) or commas (,) when referring to money (and only if necessary).
- Provide context to numbers (e.g., "1 in 20 people have…").
- Consider adding spacing for long sequences or "chunking" (grouping sequences of information into larger, more manageable units or "chunks" e.g., group a series of numbers in sets of three or four—as with this phone number: 770-777-9311).

Designing Communications for Individuals with Vision, Hearing, or Speech Impairments

As of February 2024, the World Health Organization (WHO) projected that by 2050 nearly 2.5 billion people (~1 in 10 people) will have some degree of hearing loss. And (as of 2024), the WHO estimates that globally, at least 2.2 billion people have a near or distance vision impairment (WHO, 2023). Given the prevalence of these impairments, it is prudent for ergonomics practitioners to be familiar with the guidance available specifically for persons with these impairments. Numerous references, often from regulatory bodies, outline requirements for communications, including the provision of auxiliary aids under the U.S.'s Americans with Disabilities Act (U.S. Department of Justice, 2020). For individuals with vision impairments, this includes:

- Qualified readers who can read effectively, accurately, and impartially, using any necessary specialized vocabulary.
- Information is provided in large print, Braille, or made accessible through screen-reading programs.
- Audio recordings of printed materials.

For those who are Deaf or have hearing impairments, it is important to provide options such as:

- Qualified interpreters or notetakers who can interpret effectively and impartially, both receptively and expressively.
- Transcription services and captioning devices (closed, open, or real-time).
- Written materials or printed scripts.
- Assistive listening devices and systems.

For individuals with speech impairments, options include:

- Qualified speech-to-speech transliterators who can clearly repeat unclear speech.
- Communication aids like pen and paper, communication boards, or other assistive devices, and allowing additional time for exchanges.

Audible alarms used in emergency communications, such as evacuation plans, are typically governed by regulatory standards and should accommodate Deaf or hearing-impaired individuals. For example, Occupational Safety and Health Administration's Employee Alarm Systems standard 29 CFR 1910.165 mandates that alarm systems must provide warnings that can be perceived above ambient noise and recommends adding strobe lights or tactile devices for those with hearing loss (OSHA, 2005).

While this guidance primarily addresses legally required accommodations for qualified disabilities (as defined by the ADA), these recommendations are valuable for developing inclusive communication plans overall.

Other Guidance

Whether designing a job accommodation for a specific individual's need based on their disability/impairment or designing for all to enhance the user experience for a diverse audience, other communications design guidance includes:

- Whether through visual or auditory communication channels, ensure the use of plain, clear language, avoiding slang, acronyms, or technical terms that may cause confusion or misunderstanding. This includes ensuring the style and tone suit the preferences, needs, and expectations of the intended audience, taking into account their backgrounds, experiences, and beliefs. The U.S.'s Federal Plain Language Guidelines, set forth due to the Plain Writing Act of 2010 which was enacted to enhance citizen access to government information and services (Plain Language Action and Information Network (PLAIN), 2011), includes guidance such as:
  - Writing for your audience.
  - Organizing the information.
  - Choosing your words carefully.
  - Being concise.
  - Keeping it conversational.
  - Design for reading.
  - Following web standards.
  - Testing your assumptions.
- Offer translations or interpretations of critical messages/instructions in the languages spoken by your workforce. Depending on your location and the scope of your communications—whether global or regional—there may be legal obligations to provide accurate translations. For instance, in Quebec, CA, as per Bill 96, businesses may be required to ensure communications are available in Quebecois French (Government of Quebec, 2022).
- Use visual or audible aids, such as pictures, symbols, signs, or videos, to supplement verbal or written communications, especially for those who have low literacy levels, visual impairments, or learning disabilities. The Sophia Organization (Sophia, n.d.) provides the following tips for using visuals:
  - They are easily seen or heard by the audience—e.g., during a presentation, ensure that everyone can see/hear the aids (such as those in the back of the room).

- They are easily handled, ensuring that any aid is seamlessly incorporated into any presentation.
- They are aesthetically pleasing—not distracting.
- They are easy to understand.
- They supplement the presentation—not intended to replace the presentation.

- Consider using multiple mediums for your communication messages, meaning, do not rely on just one method to convey information. Instead, incorporate a variety of multimedia channels to reach your audience more effectively. For example:
  - Visual Mediums: Use posters placed in common areas where employees gather, such as break rooms or near entrances. Digital displays can also showcase important messages, while infographics can present complex information in a more easily digestible format.
  - Auditory Mediums: Implement announcements over intercom systems or play reminder recordings during team meetings.
  - Tactile Mediums: Tactile communication methods could include raised symbols and textured surfaces to further convey important messages, such as awkward posture alerts seen in wearable technology.
  - Multimedia: Combine visual and auditory elements, such as videos that are broadcast on a loop in high-traffic areas (ideal for ergonomics training or updates).
  - Interactive Methods: Consider using QR codes on posters that link to detailed safety/ergonomics protocols or training materials online, allowing employees to engage further with the content.
- In the case of primarily verbal-based communications (National Disability Authority, 2023):
  - Begin with the purpose of the conversation.
  - Speak clearly.
  - Speak slowly.
  - Listen and allow time for others to respond; avoid interrupting or finishing other people's sentences (particularly individuals with speech impairments).
  - Keep background noise to a minimum.
  - For individuals with cognitive disabilities, use plain words and short sentences.
- Ensure communications are compatible with assistive devices, such as:
  - Using formats that assistive technologies can easily read, e.g., using plain text, HTML, or accessible PDF formats that support screen readers.
  - Using readable fonts (sans serif) with good contrast with the background.
- Facilitating open dialogue to solicit input from workers on the relevance of communications (especially if specific accommodations are initiated to meet individual needs), as well as other concerns and suggestions regarding the ergonomics process.

Use of Technology to Make Communications More Inclusive

Assistive devices and services for individuals with vision, hearing, and speech impairments, such as closed captioning, can benefit everyone. Beyond these devices, technology designed to enhance workplace ergonomics is constantly evolving. In Chapter 6, we examined various innovative tools, including virtual and augmented reality, predictive analytics using big data, wearable technology, and artificial intelligence (AI). These technologies are essential for improving risk management and creating tailored solutions for a diverse workforce within ergonomics processes.

However, it is not solely about risk management; technology also significantly enhances communication, especially when targeting a diverse team. While the earlier recommendations are valuable, they can sometimes feel overwhelming. AI, in particular, has a significant impact in this area. For instance, large language models can analyze and simplify complex texts, making dense information more accessible. Specialized writing tools can assess readability, grammar, and tone, which is especially helpful for translated materials.

Collaboration tools are another game-changer, allowing teams to review large documents together. They can analyze content, extract key information, and create concise summaries, all while enabling real-time collaboration among colleagues, including both employees and ergonomics team members. This ensures that everyone can evaluate relevance and effectiveness as discussions progress.

The industry is also witnessing a rise in ergonomics and safety management tools equipped with AI features. These tools can be used to help ensure compliance with industry-specific regulations, including communication standards like ISO 45001, which mandate effective communications in safety management systems, and ISO 7010, which governs the effectiveness of safety messages. By integrating these technologies, we can streamline ergonomics processes and enhance communication and collaboration across diverse teams.

While addressing the varied needs of a diverse workforce and developing effective communications to improve the user experience may seem daunting, technology can help to simplify this task. We just have to be intentional about learning and utilizing the available tools to enhance inclusion in our communications.

FURTHER READING

When considering disabilities or impairments, thoughts often gravitate toward mobility challenges, particularly those necessitating wheelchair use. We've extensively discussed designing for individuals with physical disabilities as well as other types of impairments, including those that could be viewed as invisible disabilities such as cognitive disabilities or neurodivergent conditions, as they all can impact how a person obtains and perceives information. It is worth noting that this book doesn't cover every possible disability, impairment, or neurodivergent condition—that would be an insurmountable task (resulting in an enormous volume, for one!). The goal of this book is to initiate conversations on how ergonomics practitioners can and should account for various impairments in the design of products, equipment, processes, and programs, and their associated communications, thereby enhancing our inclusive thinking in all aspects of our work.

And while the focus of this book and the guidance shared within has been on industrial and office settings, similar guidance (and probably more abundant) is available for digital platforms where the structure and design are critical for broad inclusion. Guidance such as (for further reading):

- World Wide Web Consortium (W3C)'s "Making Content Usable for People with Cognitive and Learning Disabilities" (W3C, 2021).
- Facil-iti.com—Web inclusion app: https://www.facil-iti.com/.
- Penn State University Accessibility & Usability website: accessibility.psu.edu ; specifically for designing for color blindness: accessibility.psu.edu/accommodations/audience/colorblindness/.
- Accessibility.com (blogs and other resources).
- U.S.'s General Service Administration Section 508's "Universal Design and Accessibility" (GSA, 2023).

REFERENCES

British Dyslexia Association. (n.d.). About dyslexia. *British Dyslexia Association.* Retrieved November 24, 2024, from https://www.bdadyslexia.org.uk/dyslexia/about-dyslexia

Cios233. (2022, July 9). Design theory lectures – color theory. *Cios233.* https://cios233.community.uaf.edu/design-theory-lectures/color-theory/

Government of Quebec. (2022). An act respecting French, the official and common language of Quebec (SQ 2022 c. 14). https://www.canlii.org/en/qc/laws/astat/sq-2022-c-14/latest/sq-2022-c-14.pdf

Ishihara, K., Ishihara, S., Nagamachi, M., Hiramatsu, S., & Osaki, H. (2001). Age-related decline in color perception and difficulties with daily activities–measurement, questionnaire, optical and computer-graphics simulation studies. *International Journal of Industrial Ergonomics, 28*(3–4), 153–163. https://doi.org/10.1016/S0169-8141(01)00028-2

Light Adviser. (2024, August 13). 5 common mistakes to steer clear of for best practices in color light usage and design. *Light Adviser.* https://lightadviser.com/5-common-mistakes-to-steer-clear-of-for-best-practices-in-color-light-usage-and-design/

Lindell, M. K., & Perry, R. W. (2012). The protective action decision model: Theoretical modifications and additional evidence. *Risk Analysis: An International Journal, 32*(4), 616–632. https://doi.org/10.1111/j.1539-6924.2011.01768.x

McFayden, J., & Malik, R. (2022, November 28). Designing for people with dyscalculia and low numeracy. *Design Notes.* https://designnotes.blog.gov.uk/2022/11/28/designing-for-people-with-dyscalculia-and-low-numeracy/

Mulholland, B. (2020, August 20). Traffic lights and UX: Smart color usage for design. *Usability Geek.* https://usabilitygeek.com/traffic-lights-ux-smart-color/#:~:text=Green%20is%20ever-positive%2C%20spurring%20you%20on%20through%20the,correct%3B%20a%20positive%20final%20action%2C%20if%20you%20will

National Center for Learning Disabilities (NCLD). (n.d.). Specific learning disabilities. *National Center for Learning Disabilities.* Retrieved November 24, 2024, from https://ncld.org/join-the-movement/understand-the-issues/specific-learning-disabilities/

National Disability Authority. (2023). Customer communications toolkit. *National Disability Authority.* https://universaldesign.ie/uploads/publications/Customer-Communications-Toolkit-2023-PDF_Final.pdf

Occupational Safety and Health Administration. (2005, July 7). Innovative workplace safety accommodations for hearing-impaired workers. *U.S. Department of Labor.* https://www.osha.gov/sites/default/files/publications/shib072205.pdf

Penn State Accessibility. (n.d.). *Bright colors and accessibility.* Penn State University. Retrieved November 24, 2024, from https://accessibility.psu.edu/color/brightcolors/

Plain Language Action and Information Network. (2011). Federal plain language guidelines. https://www.plainlanguage.gov/media/FederalPLGuidelines.pdf

Sharma, K. (2023, September 28). Balancing color consistency and effective error signaling in UI design. Medium. https://medium.com/design-bootcamp/navigating-color-consistency-and-error-signaling-in-ui-design-bf435b724028

Sophia. (n.d.)). The dos and don'ts of using visual aids. Sophia Learning. https://app.sophia.org/tutorials/the-dos-and-donts-of-using-visual-aids

Todorov, A., Chaiken, S., & Henderson, M. D. (2002). The heuristic-systematic model of social information processing. In J. P. Dillard & M. Pfau (Eds.), *The persuasion handbook: Developments in theory and practice* (pp. 195–211). Sage Publications.

U.S. Department of Justice. (2020, February 28). *Effective communication.* Americans with Disabilities Act. https://www.ada.gov/resources/effective-communication/

W3C Cognitive Accessibility Task Force. (2021, April 29). *COGA usable: Guidelines for cognitive accessibility.* World Wide Web Consortium. https://www.w3.org/TR/2021/NOTE-coga-usable-20210429/

Wogalter, M. S. (2006a). Communication-Human Information Processing (C-HIP) Model. In M. S. Wogalter (Ed.), *Handbook of warnings* (pp. 51–61). Lawrence Erlbaum Associates.

World Health Organization. (2023, August 10). *Blindness and visual impairment.* World Health Organization. https://www.who.int/news-room/fact-sheets/detail/blindness-and-visual-impairment

Yang, Z. J., Aloe, A. M., & Feeley, T. H. (2014). Risk information seeking and processing model: A meta-analysis. *Journal of Communication, 64*(1), 20–41. https://doi.org/10.1111/jcom.12082

Part III

The Current and Future State of DEI

Throughout this book, I have emphasized (perhaps to the point of repetition!) the immense value our ergonomics profession brings to Diversity, Equity, and Inclusion (DEI)—and to other areas of an organization, for that matter. In Part I, I highlighted how DEI is woven into everything we do within ergonomics. It is truly part of our DNA!

We see it in our role as designers of work—whether it is processes, products, or environments—where we are dedicated to designing for all by addressing the varying capabilities and needs of end users. We see it in our ergonomics practices, where diversity is central to our teams and initiatives, as evidenced by our participatory approach. We see it in the breadth of our profession, which encompasses a wide range of factors, recognizing that the human worker is a complex system shaped not only by their physical environment but also by cognitive, psychological, and psychosocial aspects. This allows us to address the full spectrum of human diversity. We also see it within our own field, which includes a wide variety of experts—engineers, psychologists, physical therapists, safety specialists, and more—each contributing their unique knowledge and insights, enhancing our ability to assess and support the diverse needs of workers.

In fact, we even see it in how we present ourselves to the world. Our profession's identity itself reflects diversity. We are ergonomists, human factors engineers, user experience designers, and other titles, each emphasizing different aspects of our work, but all working toward the same goal: creating inclusive, effective solutions that meet the diverse needs of the workforces we support.

DOI: 10.1201/9781032658988-16

WHAT'S TO COME?

In Part III, I will shift focus to explore the inner workings of DEI efforts and strategies. Drawing on insights from experts who have dedicated their careers to advancing DEI in various capacities, I will delve into the complexities of developing and implementing DEI strategies in today's industry landscape. This includes examining the challenges DEI faces, the predicted changes stemming from those challenges, and the opportunities for both the ergonomics and DEI domains.

As I worked on this book, I found myself grappling with an existential question: *Is this the right time for a book like this? Should I tweak the title or make other adjustments?* I pondered these questions deeply. Given the constant stream of negative headlines surrounding DEI, perhaps now is exactly the right moment for a book like this—one that aims to spark positive change, or at the very least foster lively, insightful discussions inside (and hopefully outside) our profession. To aid in this, in Chapter 13, "The State of DEI Today," I not only delve into the history of DEI and its evolution into a business strategy but also explore the current challenges facing DEI, especially in light of the rapidly changing political landscape. These shifts in laws, policies, and public perceptions are having a significant impact on how DEI initiatives are shaped and executed.

Despite these challenges, opportunities are emerging from this turbulent period, and I want our profession to be at the forefront. Chapter 14, "Using Ergonomics for Sustainable DEI Growth" explores these opportunities, focusing on how to effectively integrate DEI initiatives into today's organizational structures. The chapter highlights how ergonomics is well-positioned to support DEI, especially as organizations adapt to meet new business and societal demands. As businesses evolve, ergonomics practitioners must remain actively involved to influence organizational DEI structure changes and goals in alignment with our ergonomics objectives. Aligning ergonomics with DEI initiatives will elevate both domains from being "nice-to-haves" to essential components of any organization, particularly given the ongoing evolution of work, workforce demographics, and workspaces.

Earlier in the book, I mentioned the growing importance of the term "belonging" in the DEI lexicon. This concept has become a catalyst for progress in key performance indicators like employee satisfaction and ambassadorship, often measured by employee net promoter scores. This concept is highly relevant to ergonomics, as diverse perspectives and voices, when integrated into the ergonomics process, can influence these critical business metrics. Chapter 15, "The Wonderful World of Metrics" concludes Part III by exploring how specific ergonomics data can serve as valuable indicators for assessing DEI progress and effectiveness, especially when those metrics are tied to business outcomes. The chapter also discusses how incorporating DEI-related metrics into ergonomics processes—such as data on accommodation effectiveness, employee experience, and disability inclusion—can help organizations identify disparity gaps and develop targeted solutions to address them.

The goal of Part III is to provide ergonomics practitioners and others interested in the field with a deeper understanding of DEI strategies, demonstrating how ergonomics and DEI are not just complementary but can merge into a powerful tool for organizations. By showcasing tangible examples and strategies, Part III aims to show how integrating DEI with ergonomics can improve organizational effectiveness and drive meaningful, sustainable change.

13 The State of DEI Today

HISTORY OF DEI

Before getting to where we are today with Diversity, Equity, and Inclusion (DEI) (and the opportunity ergonomics can bring to the table), let's first delve into how we got here. In its basic form, DEI is a framework that promotes fairness of opportunity and representation through the combination of the three concepts—diversity, equity, and inclusion—to comprehensively respond to and address systemic inequity (Kaiyo et al., 2024). The Civil Rights Movement in the U.S. in the 1960s did much to spur on discussion and both societal and legislative action around the acceptance of diversity—more specifically the equal employment opportunity mandate of Title VII of the 1964 Civil Rights Act. But it was in the latter part of the 21st century where the movement to understand diversity and manage equity and inclusion within the workplace took stride, especially within the U.S. (Russen & Madera, 2021). This included the expansion of embracing a broader spectrum of diversity beyond racial diversity and including other differences within the workplace such as physical and mental abilities, ethnicity, religious, and identity groups. The increased focus on managing diversity (and the equitable treatment around it) in the workplace led to the development and implementation of structured policies and guidelines referred to as "Diversity Management" or DM. While DM policies and practices differ, most are mainly focused on unequal opportunities through activities focused on recruitment, training and employee retention, managerial accountability, and employee support. Given the significant emphasis on these activities as well as compliance, DM was and still is often the purview of human resources departments.

Similar movements to focus on equal treatment within the workplace were seen globally as well, with various countries instituting directives, such as the European Union's EU Directive 2000/43/EC Art 13, established in an effort to set forth principles of "equal treatment between persons irrespective of racial or ethnic origin" (E.U., 2000). Yet, the level of DM-related policies implemented varies from country to country—each requiring a level of geographical nuance to address specific local challenges. As the demographics of the workforce across the globe continue to evolve, so does the scope of associated policies, practices, regulations, and general DM approaches. And quite frankly, the need—as diversity is a reality that businesses must account for or risk long-term sustainability.

ATTITUDE SHIFT ABOUT DEI

We can't have a book about DEI without laying it out bare. It is no secret that the discussion around and about "DEI" can be challenging, especially as of late. The

DOI: 10.1201/9781032658988-17

words themselves have led to heated discussions and debates, simply on semantics alone. With the terms strongly associated by some as a form of political activism, the DEI movement has undergone rebranding in some circles as a means of trying to avoid controversy and better navigate the political landscape while continuing to initiate positive change.

Recent social unrest events and subsequent calls for justice seen around the world (e.g., the murder of George Floyd and the #MeToo movement in the U.S.; the death of Mahsa Amini leading to protests in Iran, and the Black Lives Matter movement that rippled across the globe), as well as the COVID-19 pandemic that highlighted a call to focus on neurodiversity, have further ignited the discussion around DEI, especially its role within the workplace and how that trickles out into society as a whole. However, and possibly due to the countering of these events, DEI has faced significant challenges. For example, it has been shown that when businesses are faced with adversity (such as economic uncertainty), DEI initiatives are often among the first things to be cut in the name of optimization and cost-cutting (Pringle & Strachan, 2015). Sadly, it has also been seen that generational changes that evoke significant societal change and conflict can lead to intolerance becoming more accepted, even in the face of rhetorical commitments to equity (Haack & Sieweke, 2018). On top of these attitudes that threaten or deter DEI efforts, there are distractors as well. Within the U.S., there has been severe political backlash that has threatened the advancement of DEI efforts. From a governmental perspective, particularly within the U.S., at the time of this book being written, there are numerous "anti-DEI" laws making their way through federal, state, and local legislation as backlash to the push and call for social justice change. These laws, as well as the very vocal anti-DEI advocates, have caused businesses to shy away from public pronouncements of DEI commitments (Kratz, 2024b). To delve more into this, I spoke with Marenda Caldwell, a Certified Professional Ergonomist who has spent the past several years studying the varied DEI programs and initiatives across global organizations. With regards to the shift Marenda shared her observations:

> In 2020–2021, major corporations increased their focus on DEI and Social Justice initiatives. But by 2022 it felt like the momentum had faded. The plan of incorporating DEI into the core framework of the core strategy had some stickiness but as new laws and mandates are being promulgated the investment in these programs seems to also have faded.

In response to these legal challenges, there has been a noticeable increase in rebranding and renaming efforts. Terms like "Belonging" or "Culture" are now being added to or replacing the previous DEI efforts within organization (Mallick, 2024). Additionally, some initiatives have been completely repackaged under "Employee Experience" (Liu, 2024).

I had the opportunity to speak with Tiffane Davis, a DEI expert in this space with over 20 years of experience, including developing and executing brand strategies. She highlighted the significant role that her background in marketing plays in effectively communicating DEI initiatives internally within organizations and externally to the public. For Tiffane, who defines DEI as ensuring that people with diverse backgrounds, experiences, and talents are given fair opportunities to succeed and feel valued for who they are and their contributions, a significant challenge impacting attitudes toward DEI is the lack of buy-in from leadership. Additionally, companies often face a shortage of resources (such as funding, staff, training, tools, and time) and data (including metrics, benchmarks, surveys, and feedback). This lack of support and resources leads to unclear ambitions, making it difficult for companies to set clear and realistic goals, measure progress and impact, and identify areas for improvement in their DEI efforts. Without leadership buy-in, DEI initiatives may be perceived as optional, superficial, or tokenistic, rather than strategic, meaningful, and essential.

I also had the opportunity to talk to Jourdan Saunders, the founder and CEO of The Resource Key, a company that aims to use marketing as a tool for change and to embed inclusion and accessibility in the marketing space and processes (like Tiffane, Jourdan believes that marketing is a powerful way to reach every department and to promote DEI values). She shared with me her views on DEI, which are influenced by her personal and professional experiences. As a mom, wife, speech-language pathologist who has provided direct therapy for more than a decade, and a person with aphantasia—which is a condition where a person cannot visualize images in their mind that impacts approximately 2% of the population—she has a unique perspective on DEI. To Jourdan, DEI is, from a high level, a commitment to learning and understanding backgrounds, perspectives, and different environments, recognizing barriers, but also valuing and promoting peoples' strengths and diversity. To her, it is about embracing and leveraging diverse backgrounds, perspectives, and abilities, as well as understanding how to foster a more inclusive environment. She also considers how different environments can affect the DEI discussion, as things can change depending on the context.

As for specific barriers to the effective implementation of DEI efforts, we can turn to a study conducted by Aragon & Kim (2021), which included interviews with nine field experts focused on D&I initiatives within their respective organizations. For the study, questions asked included what barriers exist for diversity and inclusion in business today. The collective sentiments from those interviewed noted resistance, bias, and feelings of threat as barriers due to reasons including organizational structures, inflexible options, inequitable landscapes, and limited resources such as time and budget.

But while there is no shortage of debate around how DEI should be approached and managed (if at all) and the challenges—old and new—that surround it, it is still seen in a positive light. According to a 2023 Pew Research Study (Minkin, 2023) looking at attitudes toward DEI, the majority of U.S. adults view efforts around DEI as a "good thing." And organizations continue to recognize and tout the benefits of a DEI (Kratz, 2024a).

It remains that an inclusive workplace that recognizes, respects, and actively incorporates diversity (in all its forms) in business activities is inherently good for business. For example, as to why,

- Diversity in the workplace opens doors to many new viewpoints and expertise (Tang et al., 2017).
- Diverse teams with varying perspectives and approaches help them compete better in today's fast-changing global market (Park et al., 2023; Allison & Flaherty, 2020; Shan et al., 2017).
- Programs that focus on diversity lead to increased employee engagement and higher salesperson performance (Luu et al., 2019; Fein et al., 2023) and that engaged employees tend to show increased productivity and contribute positively to team dynamics while sharing innovative ideas (Trochmann et al., 2023).
- Diverse workplaces contribute to better business performance (Zhang, 2020; Kim et al., 2005).
- With the proper conditions, individuals and teams of different backgrounds, experiences, and different worldviews can perform better with regard to innovatively thinking outside of the box (Ng and Stephenson, 2016).
- With respect to inclusion specifically, Combs et al. (2019) and Madera et al. (2017) found that a highly inclusive environment can create enhanced information sharing, creativity, job satisfaction, psychological safety, and affective commitment.
- Organizations with strong commitments to DEI are in a better position with regard to the recruitment and retention of skilled workers (Glassdoor, 2020).
- Robust DEI strategies can improve an organization's internal and external image (Behlow, 2019; Environics International Ltd., 1999).

Despite the challenges and the mixed attitudes, the consensus from those within the industry (Basiouny, 2024), DEI, in whatever form it evolves into, is here to stay. The intent of DEI efforts is seen as valuable and good for business, hence the effort of some organizations to rebrand to keep efforts ongoing while also navigatng the legal challenges and potential contentious backlash from some.

OPPORTUNITY IN THE FACE OF CHALLENGES

Given the benefits of DEI, organizations want to continue to recognize and incorporate DEI into their way of doing business. It is those challenges and barriers described earlier surrounding DEI, though, that have organizations reconsidering and refining their DEI strategies. Unfortunately, though, companies, grappling with the complex and evolving legal landscape, are seen scaling back resources dedicated to DEI efforts, including specialized teams and initiatives (Francis & Cutter, 2024).

When I asked Jourdan about the challenges with DEI and strategies for it, she noted that many of the challenges with the view of DEI have to do with the perception that DEI can be seen as a "check the box" activity. For example, when a specific goal or milestone is met—say a certain gender percentage breakdown in a company's executive management—then efforts wane as opposed to being ongoing efforts organically embedded into the company's culture. This is also compounded, Jourdan says, by the fact that DEI tends to be separated out into its own silo. When it is its own silo and funding is dedicated to it, when funding is reduced or removed, DEI efforts unfortunately cease or are extremely limited. However, to her, DEI is a natural component of many areas and domains within the workplace, such as in innovation efforts or in communications, and recognizing that should be the conversation around DEI moving forward, especially given the ongoing funding reductions seen across the United States. She adds,

> It is really going to depend on each department when you look at it structurally. So, I think the key conversation will be something like, "Alright, the funding is going away, but the need for DEI isn't." The focus will likely be on finding ways to sustain these efforts long-term. That's where I believe much of the discussion will center—how to ensure DEI remains ongoing.

Moving forward, companies are seeking approaches that not only highlight DEI as a fundamental driver of performance improvement but also as a transformative force for cultural change. This strategic realignment presents a significant opportunity to enhance the effectiveness and efficiency of DEI efforts.

In Quinetta Roberson's article "How Integrating DEI Into Strategy Lifts Performance" (Roberson, 2024) for the Winter 2025 MIT Sloan Management Review, she notes that organizations are seeking evidence-based, data-driven DEI strategies that align with company values and can lead to competitive advantages. She further delineates the keys to achieving this, which include first integrating DEI efforts into deeper corporate objectives and goals that allow for the creation and management of inclusive, diverse, and equitable environments, as opposed to focusing on specific HR areas like recruitment, training, or retention. The second key Ms Roberson noted is to implement DEI practices that collaborate with other areas of the company to work together to achieve common goals and expand their effects. And lastly, organizations should focus on developing tools to measure and evaluate change through DEI.

In my conversation with Marenda, from her experience and exploration in this space, she sees organizations increasingly aligning their DEI strategies with their overall corporate strategies. She adds,

> In my experience there are generally three core components: People/Talent, Business Outcomes/Innovation, and Customer focused.

According to her, these strategies are primarily driven by senior leadership and supported by a governing body, as well as employee-driven business resource groups or affinity groups (to be discussed in more detail in the following chapter). A key aspect of this structure, she notes, is that these employee groups form the foundation that supports *all* pillars of the corporate strategy, not just DEI.

What does this mean for us as ergonomics practitioners? The evolution in the execution of DEI strategies presents a remarkable opportunity for the field of ergonomics and its associated processes within organizations. The argument of this book is not that ergonomics replaces DEI as a corporate strategy, but that it should be closely integrated with current and future DEI efforts. By leveraging the inherent qualities of our processes and the field in general, which align with the suggested keys for sustainable DEI advancement and corporate culture growth, we can enhance the impact of DEI initiatives. This alignment will enable DEI and its collective practices to be recognized as essential tools for driving business performance and contributing to sustainable value creation for the organization. And by extension, ergonomics as well.

Conversely, ergonomics can also take advantage of existing frameworks and initiatives that promote DEI, such as those associated with environmental, social, and governance, corporate social responsibility, climate action, sustainability, and the Sustainable Development Goals. These areas are all relevant to DEI and are areas where ergonomics can provide valuable insights, solutions, and highly promotable case studies—such as in-depth examples of inclusive work design projects that reduced occupational safety risks, enhanced employee well-being, increased gender diversity, and improved process efficiency while reducing CO_2 waste—that benefit both the organization and its stakeholders. By strategically aligning ergonomics with these broader agendas and providing ready-made examples that demonstrate the work we're doing is advantageous to those goals, we can enhance the visibility and value of ergonomics. This alignment can also potentially attract more funding and support to ergonomics efforts as organizations do so in turn from investors, partners, and customers who seek organizations committed to social initiatives.

This is a critical move, Jourdan notes, because how businesses are being funded is starting to shift. And to continue to advance DEI efforts, as well as ergonomics efforts, it would be key to show how both can play in these spaces.

REFERENCES

Allison, L., & Flaherty, K. E. (2020). Investigating firm-level drivers of salesperson brand identification. *Journal of Business Research, 121*, 154–169. https://doi.org/10.1016/j.jbusres.2020.08.030

Aragon, A., & Kim, S. S. (2021). Drivers and barriers of diversity and inclusion in business: Interviews with nine field experts. *DU Undergraduate Research Journal Archive, 2*(1), Article 5. https://digitalcommons.du.edu/duurj/vol2/iss1/5

Basiouny, A. (2024, June 10). *Is DEI going away? Here's what experts say. Knowledge at Wharton.* University of Pennsylvania. https://knowledge.wharton.upenn.edu/article/is-dei-going-away-heres-what-experts-say/

Behlow, G. (2019, September 12). African American spending power demands that marketers show more love and support for Black culture. *Nielsen.* https://www.nielsen.com/news-center/2019/african-american-spending-power-demands-that-marketers-show-more-love-and-support-for-black-culture/

Combs, G. M., Milosevic, I., & Bilimoria, D. (2019). Introduction to the special topic forum: Critical discourse: Envisioning the place and future of diversity and inclusion in organizations. *Journal of Leadership and Organizational Studies, 26*(3), 277–286. https://doi.org/10.1177/1548051819843797

Environics International Ltd. (1999, December). *The millennium poll on corporate social responsibility: Global public opinion on the changing role of companies.* Environics International Ltd. https://docslib.org/doc/9351111/the-millennium-poll-on-corporate-social-responsibility

European Union. (2000). Directive 2000/43/EC of the European Parliament and of the Council of 29 June 2000 implementing the principle of equal treatment between persons irrespective of racial or ethnic origin. *EUR-Lex.* https://eur-lex.europa.eu/legal-content/EN/TXT/PDF/?uri=CELEX:32000L0043

Fein, E. C., Tziner, A., & Vasiliu, C. (2023). Perceptions of ethical climate and organisational justice as antecedents to employee performance: The mediating role of employees' attributions of leader effectiveness. *European Management Journal, 41*(1), 114–124. https://doi.org/10.1016/j.emj.2021.11.003

Francis, T., & Cutter, C. (2024, December 28). Corporate America drew back from DEI. The upheaval isn't over. *The Wall Street Journal.* https://www.wsj.com/business/corporate-dei-policy-changes-pull-back-33fdab87

Glassdoor. (2020, October 1). Glassdoor's diversity and inclusion workplace survey. *Glassdoor.* https://www.glassdoor.com/blog/glassdoors-diversity-and-inclusion-workplace-survey/

Haack, P., & Sieweke, J. (2018). The legitimacy of inequality: Integrating the perspectives of system justification and social judgment. Journal of Management Studies, 55(3), 486–516. https://doi.org/10.1111/joms.12323

Kaiyo, A. N., Gande, M., Murapa, K. A., & Mupa, M. N. (2024). Unmet standards for Diversity, Equity, and Inclusion (DEI) in the USA & recommendations to meet the standards. *Iconic Research and Engineering Journals, 8*(4), 499–513.

Kim, W. G., Leong, J. K., & Lee, Y. K. (2005). Effect of service orientation on job satisfaction, organisational commitment, and intention of leaving in a casual dining chain restaurant. *International Journal of Hospitality Management, 24*(2), 171–193. https://doi.org/10.1016/j.ijhm.2004.05.004

Kratz, J. (2024a, March 5). Are companies quiet quitting DEI? *Forbes.* https://www.forbes.com/sites/juliekratz/2024/03/05/are-companies-quiet-quitting-dei/

Kratz, J. (2024b, June 9). Anti-DEI legislation: How we got here with tips to combat it. *Forbes.* https://www.forbes.com/sites/juliekratz/2024/06/09/anti-dei-legislation-how-we-got-here-with-tips-to-combat-it/

Liu, X. (2024, January 3). *The integration of design, DEI, and employee experience* [LinkedIn post]. LinkedIn. https://www.linkedin.com/pulse/integration-design-dei-employee-experinece-xi-liu-shrm-scp/

Luu, T. T., Rowley, C., & Vo, T. T. (2019). Addressing employee diversity to foster their work engagement. *Journal of Business Research, 95*, 303–315. https://doi.org/10.1016/j.jbusres.2018.08.017

Madera, J. M., Dawson, M., & Neal, J. A. (2017). Managers' psychological diversity climate and fairness: The utility and importance of diversity management in the hospitality industry. *Journal of Human Resources in Hospitality and Tourism, 16*(3), 288–307. https://doi.org/10.1080/15332845.2017.1308987

Mallick, M. (2024, December 27). *This is how DEI will change in 2025.* Fast Company. https://www.fastcompany.com/91247655/this-is-how-dei-will-change-in-2025

Minkin, R. (2023, May 17). *Diversity, equity, and inclusion in the workplace.* Pew Research Center. https://www.pewresearch.org/social-trends/2023/05/17/diversity-equity-and-inclusion-in-the-workplace/

Ng, E., & Stephenson, J. (2016). Individual, teams, and organizational benefits of managing diversity: An evidence-based perspective. In A. M. Konrad, P. Prasad, & J. K. Pringle (Eds.), *The Oxford handbook of diversity in organizations* (Chapter 11). Oxford University Press.

Park, Y. W., Voss, G. B., & Voss, Z. G. (2023). Advancing customer diversity, equity, and inclusion: Measurement, stakeholder influence, and the role of marketing. *Journal of the Academy of Marketing Science, 51*(1), 174–197. https://doi.org/10.1007/s11747-022-00883-6

Pringle, J. K., & Strachan, G. (2015). Duelling dualisms: A history of diversity management. In R. Bendl, I. Bleijenbergh, E. Henttonen, & A. J. Mills (Eds.), The Oxford handbook of diversity in organizations (pp. 39–59). Oxford University Press.

Roberson, Q. (2024, November 7). *How integrating DEI into strategy lifts performance.* MIT Sloan Management Review. https://sloanreview.mit.edu/article/how-integrating-dei-into-strategy-lifts-performance/

Russen, M., Dawson, M., & Madera, J. M. (2021). Gender diversity in hospitality and tourism top management teams: A systematic review of the last 10 years. *International Journal of Hospitality Management, 95*, 102898. https://doi.org/10.1016/j.ijhm.2021.102898

Shan, L., Fu, S., & Zheng, L. (2017). Corporate sexual equality and firm performance. *Strategic Management Journal, 38*(9), 1812–1826. https://doi.org/10.1002/smj.2624

Tang, G., Yu, B., Cooke, F. L., & Chen, Y. (2017). High-performance work system and employee creativity: The roles of perceived organisational support and devolved management. *Personnel Review, 46*(7), 1318–1334. https://doi.org/10.1108/PR-09-2016-0235

Trochmann, M., Stewart, K., & Ragusa, J. (2023). The impact of employee perceptions of inclusion in a racially diverse agency: Lessons from a state government survey. *Public Personnel Management, 52*(4), 543–565. https://doi.org/10.1177/00910260231187544

Zhang, L. (2020). An institutional approach to gender diversity and firm performance. *Organisation Science, 31*(2), 439–457. https://doi.org/10.1287/orsc.2019.1297

14 Using Ergonomics for Sustainable DEI Growth

Given the well-documented benefits of ergonomics in enhancing occupational safety and operational efficiency, it is an essential element within an organization's structure. Regardless of its placement within the organizational chart, the growing trend of integrating Diversity, Equity, and Inclusion (DEI) into existing frameworks—rather than treating it as a standalone initiative while maintaining C-suite level management and accountability—presents a significant opportunity for ergonomics to add substantial value. The core message of this book is that by aligning ergonomics efforts with DEI goals, and vice versa, organizations can create more inclusive and supportive environments that address the diverse needs of all employees, thereby fostering better worker well-being and enhanced operational efficiency. The synergy between ergonomics and DEI amplifies their effectiveness, creating a more holistic approach to workplace inclusion, which can potentially transform company culture, the ultimate goal for both areas.

But How?
Ergonomics processes inherently embody many of the values and principles crucial for advancing DEI goals. This chapter will explore how organizations can strategically leverage the values inherent in ergonomics—such as inclusion, equity, and accessibility—to support DEI strategies. Specifically, we will discuss how ergonomics can play a pivotal role in helping organizations meet the key principles for sustaining DEI efforts in this evolving and, in some respects, challenging climate.

ERGONOMICS AND THE FIRST AND SECOND KEYS OF DEI STRATEGIES

Established ergonomics processes, which may vary from organization to organization, essentially embody the strongly supported Values-Principles Model (VPM) approach proposed by MIT Sloan for advancing DEI objectives (Beach and Segars, 2022). This validated model provides organizations with both the blueprint for implementation—which can be applied at the individual, organizational, and societal levels—and a means to measure the aspects of DEI efforts within the organization.

Ergonomics and the VPM Values

The proposed VPM consists of four values: Representation, participation, application, and appreciation. When these values are integrated into a system, they can

DOI: 10.1201/9781032658988-18

create a continuously engaged process. The alignment of our ergonomics process with these values makes our collaboration with DEI immediately beneficial, especially as new DEI strategies are being considered. Below, I outline how I see ergonomics processes contributing to, and often aligning with, the intent of each value.

- **Representation**: Ergonomics truly embodies representation by recognizing and addressing diverse differences in risk assessments—whether physical, cognitive, behavioral, or social. This approach goes beyond merely identifying and mitigating risks; it directly influences the design of work environments and tasks to enhance both safety and operational efficiency. By tailoring work processes to meet the unique needs of individuals, considering their physical abilities, cognitive styles, behavioral patterns, and social contexts, ergonomics not only improves safety and performance but also empowers individuals to feel capable and confident in their roles. This personalized approach fosters a deep sense of representation, contributing to feelings of belonging and inclusion among employees.

While meeting the broad spectrum of needs can pose challenges, it also brings opportunities for innovation and creativity. By embracing diversity, organizations can discover new solutions and approaches to everyday tasks. Supporting an ergonomics process that actively creates environments where all employees feel seen, respected, and included sends a clear, visible message of organizational commitment to valuing each individual. This approach reflects a company's dedication to respecting the uniqueness of its employees, directly impacting their sense of well-being, engagement, and overall contribution to the business's success.

- **Participation**: The ergonomics process is fundamentally intertwined with the value of participation, especially when adopting a Participatory Ergonomics approach. This approach has been characterized as a philosophy that centers around the inclusion of workers by equipping them with the tools, knowledge, and authority in the form of decision-making power and resources to influence meaningful change in the planning and organization of their own work (Liker et al., 1989). I should point out that the last point of "authority" implies a level of accountability that is placed on the organization in relation to its commitment to ergonomics, and by extension, DEI efforts. The level of engagement in this approach can vary across different organizations, but even at its most basic level, an ergonomics process relies heavily on employee involvement to be successful. At a minimum, this participation is essential for gathering input for risk assessments and designing user-centered solutions. However, the true value of participation is realized at higher levels of engagement, where employees actively contribute their experiences and perspectives. This deeper involvement ensures that representation becomes meaningful, as mere representation without active participation holds little value.

The structure of the ergonomics process itself also plays a crucial role in creating an environment that supports inclusive and effective teamwork, as well as creative innovation. By encouraging open communication, fostering diverse input, and focusing on employee well-being, the ergonomics process becomes a tool not only for individual comfort but also for enhancing organizational collaboration and inclusion. This approach helps create a workplace where everyone feels seen and included, demonstrating the organization's respect and value for the individuality of its employees.

- **Application**: A key aspect of the "application" value is the focus on human-centeredness in organizational processes. Ergonomics, by definition, is inherently human-centered, aiming to design work systems that balance worker capabilities with system requirements to ensure effectiveness.

Another key aspect of this value is the challenge to entrenched systems that may disproportionately benefit certain groups. The ergonomics process identifies and redefines work processes and environments that may have previously been optimized for only a specific group of workers (e.g., male-dominated industries and tasks), potentially excluding others. This challenge to the status quo drives efforts to improve workplace systems, expanding the pool of individuals who can perform tasks and enhancing safety and efficiency for everyone.

Ergonomics also emphasizes the importance of applying human-centered principles beyond the workplace. The inclusive design principles of ergonomics extend to the products and services offered by an organization, ensuring they meet the needs of a diverse customer base and align with the applied values of inclusion and accessibility.

Lastly, as each organization's DEI journey should be tailored to their specific needs and culture, incorporating that journey with that organization's ergonomics process helps to ensure that is the case, as the ergonomics process is inherently tailored to the specifics of the workplace, the workforce, and work processes. As such, this collaboration allows for a holistic assessment of how different individuals interact with their work environment, ensuring that DEI efforts are not just an add-on but are embedded into the core of workplace design and operations, and thus, the organization's culture.

- **Appreciation**: Ergonomics directly supports this value through its participatory approach, emphasizing connectedness and inclusion in how employees engage with the process and collaborate to design, assess, and improve work systems. Work is inherently a collective process, and decisions made at any stage, whether upstream or downstream, can significantly impact the overall system. This interconnectedness is a central tenet of the ergonomics process, illustrating the importance of inclusion, collaboration, and appreciation for the actions employees take to support and assist one another.

With robust ergonomics programs and processes, these contributions are often recognized and celebrated across the organization and externally (e.g., Applied

Ergonomics Conference Ergo Cup competition). In fact, it is this celebratory attitude that I think sets ergonomics apart (I mean, if you can't have fun, why do it?). Through these celebratory events, best practices and employee-driven innovations are frequently showcased, highlighting the importance of individual input and reinforcing the value of collective efforts. This recognition plays a key role in the sustainability of ergonomics initiatives, motivating employees to continue engaging in the process and reinforcing the idea that all contributions, regardless of their size or scope, are valued. By emphasizing appreciation of employee contributions, ergonomics fosters a culture where everyone is empowered to help improve the workplace for the benefit of all.

"Equity" or "Inclusion" First

Another reason ergonomics and an ergonomics process can effectively drive diversity and equity goals relates to how ergonomics has a direct impact on "inclusion." The evolution of the term "Diversity, Equity and Inclusion" has led to several arguments regarding the order of the terms. Initiatives specific to diversity have long been a focus for businesses implementing DEI efforts, primarily because of the sheer fact that diversity is much more visibly tangible and easier to see and measure compared to the equity and inclusion elements, which rely on first-hand experiences that can be challenging to capture (Carlini and Grace, 2021; Romansky et al., 2021; Tracy et al., 2020). As such, emphasis on diversity elements reflect heavily within traditional Diversity Management policies and practices, such as efforts around increasing and measuring diversity within recruitment, retention, and training. However, while more difficult to assess, both the equity and inclusion elements are seen as enablers of diversity metrics (Carlini & Grace, 2021; Mor Barak et al., 2021; Romansky et al., 2021). Thus, it has been proposed and argued that the primary focus be shifted to recognize how equity and inclusion efforts can serve as better drivers for diversity. As for "equity" being the initial focus—or specific organizational actions that support what is perceived as fair, Berry (2016) noted that focusing on diversity without first implementing or equally focusing on equity actions could lead to the separation or segregation of groups. Ferdman (2017) noted that equity should be applied to both diversity and inclusion to remove any barriers that would impede contribution, opportunity, and achievement.

The argument for focusing on "inclusion" initially—or specific organizational actions that reflect the acknowledgment of individuals having unique and different needs, thus there is a recognition that individuals may require different (yet equitable) treatment—is that having an inclusive culture and inclusive climate will naturally help facilitate both equity and diversity. Russen and Dawson (2023) argued that without implementing inclusionary values first and at the cultural level, implementing diversity initiatives would be futile.

As for both in the question of what should come first or be a primary focus, the intent is to still treat the approach as circular (e.g., IED or EID), because if the environment does not recognize and value differences then there is no benefit for DEI (Russen & Dawson, 2023). With that line of thought, ergonomics specifically brings tremendous value for both a focus on equity and inclusion in that the physical, psychophysical, and psychosocial environment directly relates to how a person/worker, given their specific needs, is able to interact with their environment as related to their physical, psychophysical, and psychosocial experiences and expectations ("equity") and how they feel included ("inclusion") and able to give their best and thrive. It is also supported by the VPM that when the values are deployed as a system, it enables diversity to grow (Beach & Segars, 2022). Meaning, with ergonomics, the organization will provide a work environment that is not a barrier for a diverse workforce and is able to reflect as such, particularly in their metrics.

Employee Resource Groups—DEI's Secret Weapon

Whether prioritizing "inclusion," "equity," or "diversity," without a circular approach, there is a risk that one aspect may become the sole focus, isolated from and/or at the expense of the others. Engagement is the key element that ties them all together. Throughout this book, a common theme is participation, whether discussing the ergonomics process or advancing DEI efforts. Without engagement from all involved—employees at all levels—all efforts, whether related to ergonomics or DEI, are ineffective. I have elaborated on the joys and benefits of a Participatory Ergonomics process and how empowering employees can significantly impact the sustainability of ergonomics programs and processes. This participatory concept is equally critical for the effectiveness of any DEI strategy, highlighting the importance of employee resource groups (ERGs).

These groups help bridge the gap between management and employees, ensuring that DEI strategies are not only top-down directives but also grassroots movements that reflect the true needs and aspirations of the workforce. With adequate support, resources, and guidance, they also embody the three other values—representation, application, and appreciation.

Employee resource groups—also known as business resource groups, affinity groups, as well as a whole host of names that signify a collection of people with shared interests and/or identities—originated out of the Civil Rights Movement in the U.S. In 1970, Joseph Wilson, the CEO of Xerox at the time, established the National Black Employee Caucus to advocate for inclusion and policy change in response to the racial tensions prevalent in the country (Catalino et al., 2022). From there, similar groups of employees within organizations have been established across the globe. As of late, organizations have renewed their focus on ERG initiatives due to challenges with worker disengagement and retention, the resurgence of union organizing, and a push for greater workplace inclusion (Leger, 2023). As such, ERGs have become an instrumental component of DEI strategies, especially as the approach to DEI gets a closer look by organizations.

By connecting employees from diverse backgrounds, identities, and experiences, ERGs provide a platform for sharing stories and offering mutual support. This enables employees to voice their opinions and influence the organization's DEI strategy. ERGs can identify the issues and challenges their members face and propose ideas and solutions to improve the situation for everyone. In her perspective, Jourdan adds,

I think they're at the core because that's where the people are coming together.

The structure and governance of these ERGs vary from company to company, even from group to group. But generally speaking, according to author Matthew Leger, ERGs are typically employee-led, sponsored (via the allocation of funding and other resources) by company leadership, have goals that align with the organization, and are company oriented. In terms of DEI efforts, these groups can advance these efforts through several avenues, according to McKinsey & Co. For starters, ERGs can be essential for fostering inclusion in the workplace by aligning their goals with both employee expectations and corporate DEI priorities. Also, the ERGs can provide a place for employees to feel a sense of belonging and connection to a community and potentially alleviate the feeling of being the only one (in terms of race, gender, ethnicity, etc.) within the workplace.

Another way ERGs can help to advance DEI efforts, specifically diversity, is through participation in talent attraction, such as participating in career fairs and other recruiting events, thus potentially improving recruitment. Additionally, members of the ERGs have valuable connections with internal groups and can provide valuable guidance in retention efforts. Similarly, ERG members have valuable connections with external local communities, which could be helpful for outreach and helping the organization understand community nuances to aid in how they do business. Doing this can drive positive social impact through community engagement and improve brand visibility.

DEI's reliance on ERGs as a foundational component aligns very closely with the Participatory Ergonomics approach. Marenda shares her perspective,

> [Both] align on the basis of leveraging all employees to engage in identifying risks and improving working conditions. This approach increases employee acceptance and adoption of changes and thereby drives safety, productivity, and competitive improvements.

Just as recognizing and leveraging the unique perspectives and ingenuity of employees through a Participatory Ergonomics approach, tapping into ERGs is a way for organizations to recognize and make tangible benefits from diversity. However, there are challenges, namely structural, that can impede the effectiveness and sustainability of these groups. One such challenge is the potential lack of sustainable organizational structure and governance. For organizations with established, robust ergonomics/safety processes with employee-driven teams, these could be used as models for setting up local ERGs. This includes having documented rules, policies, processes, standard operating procedures, and guiding principles that clearly outline how they should operate and what activities are within and outside their scope.

Another challenge is the risk of burnout among ERG leaders and members, who are often volunteers, due to ineffective efforts. The organization's ergonomics process can provide strong collaborative support to ERGs. This support could include periodic reviews of relevant metrics (to be discussed further in Chapter 15) to identify work design gaps that need prioritization (e.g., physically demanding jobs that are disproportionately male) or collaboration on targeted projects to improve the inclusion of individuals with unique needs (e.g., designing a new office area to meet the needs of neurodiverse employees or those with mobility impairments). Ultimately, by strategically aligning ergonomics with the values and principles of DEI and collaborating with and advancing established DEI engagement structures such as ERGs, we can demonstrate that ergonomics is not just a luxury or an add-on, but a strategic and essential tool for any organization.

ERGONOMICS AND THE THIRD KEY OF DEI STRATEGIES

Speaking of metrics, in relation to the third key mentioned in the MIT Sloan Review article—developing tools to measure and evaluate change through DEI—a company's ergonomics process and the measures of success within it can also be used to measure the effectiveness of DEI initiatives and analyze progress along DEI goals. Ergonomics is a way to assess—quantitatively and qualitatively—inclusion, which has traditionally been a challenge with current Diversity Management.

Having meaningful and applicable metrics that reflect the key elements of DEI efforts speaks to the accountability of the organization. Additionally, these ergonomics-related metrics can help to provide the data-driven evidence needed to make informed decisions on DEI direction and strategy, and ultimately make DEI less compliance-driven and transactional (as it has been traditionally seen as (Beach and Segars, 2022) and more culture-transforming. These will be explored further in the following chapter.

REFERENCES

Beach, A., & Segars, A. (2022). How a values-based approach advances DEI: A new model for developing diversity, equity, and inclusion in the organization can increase employee satisfaction. *MIT Sloan Management Review*, *63*, Summer 2022 Issue.

Berry, J. (2016). Diversity and equity. *Cross Cultural and Strategic Management*, *23*(3), 413–430. https://doi.org/10.1108/CCSM-04-2015-0080

Carlini, J., & Grace, D. (2021). The Corporate Social Responsibility (CSR) internal branding model: Aligning employees' CSR awareness, knowledge, and experience to deliver positive employee performance outcomes. *Journal of Marketing Management*, *37*(7–8), 732–760. https://doi.org/10.1080/0267257X.2021.1911966

Catalino, N., Gardner, N., Goldstein, D., & Wong, J. (2022, December 7). *Effective employee resource groups are key to inclusion at work. Here's how to get them right.* McKinsey & Co. https://www.mckinsey.com/capabilities/people-and-organizational-performance/our-insights/effective-employee-resource-groups-are-key-to-inclusion-at-work-heres-how-to-get-them-right

Ferdman, B. M. (2017). Paradoxes of inclusion: Understanding and managing the tensions of diversity and multiculturalism. *The Journal of Applied Behavioral Science*, *53*(2), 235–263. https://doi.org/10.1177/0021886317694407

Leger, M. (2023). *Building high-trust workplaces: The role of employee resource groups in fostering stronger employee-employer relationships.* American Enterprise Institute. http://www.jstor.org/stable/resrep55292

Liker, J. K., Nagamachi, M., & Lifshitz, Y. R. (1989). A comparative analysis of participatory ergonomics programs in US and Japan manufacturing plants. *International Journal of Industrial Ergonomics*, *3*(3), 185–199.

Mor Barak, M. E., Luria, G., & Brimhall, K. C. (2021). What leaders say versus what they do: Inclusive leadership, policy-practice decoupling, and the anomaly of climate for inclusion. *Group and Organization Management*, *47*(4), 840–871. https://doi.org/10.1177/10596011211007334

Romansky, L., Garrod, M., Brown, K., & Deo, K. (2021). How to measure inclusion in the workplace. *Harvard Business Review*, *27*, 1–7.

Russen, M., & Dawson, M. (2023). Which should come first? Examining diversity, equity, and inclusion. *International Journal of Contemporary Hospitality Management*, *36*, Article 1184. https://doi.org/10.1108/IJCHM-09-2022-1184

Tracy, S. J., Razzante, R. J., & Hanna, K. N. (2020). Creating the being of inclusion in organizations. In M. L. Doerfel & J. L. Gibbs (Eds.), Organizing inclusion: Moving diversity from demographics to communication processes (pp. 113–132). Routledge.

15 The Wonderful World of Metrics

As noted before, having meaningful metrics is critical for accountability, showing progress, and decision-making for Diversity, Equity, and Inclusion (DEI) efforts. What metrics are tracked can vary just as how DEI strategies can vary from company to company, and from country to country. Referring back to my conversation with Marenda, she noted that in some cases, there are governmental requirements for tracking and reporting certain diversity metrics, such as in the U.S. (with minority group reporting generally only reported for U.S. sites and gender data reported as a percentage based on self-disclosure data). However, in other countries, such tracking can be prohibited. For example, in the E.U., some data privacy and transparency requirements may limit the collection of demographic data, such as with Article 9(1) of the General Data Protection Regulation, where information such as racial or ethnic origin can be prohibited unless an exception applies under GDPR Article 9(2) (Coutelier & Engfeldt, 2022). Thus, just as DEI programs/initiatives need to be tailored to the specific country/region/state/etc. based on their population and other regional and geographical nuances, such is the case concerning policies and laws in relation to metrics and tracking.

According to the Diversity Best Practices offered by Stephen Thompson (Thompson, 2017), DEI themes that typically become important metrics include:

- Inclusion, Engagement, Productivity.
- Recruitment.
- Retention.
- Development and Advancement.
- Market Impact.
- Compliance.

Additionally, the Best Practices advise that the metrics be distinctly linked to business outcomes (as in showing impact on critical business progress), leverage the power of engagement (as in measuring effectiveness of communication, feedback mechanisms, development support, the job assignment process, other business processes, and infrastructure), and insist on accountability (as in developing standards for comparison such as with tracking against benchmarks or baselines).

DEI METRICS: BEST PRACTICES (AND HOW ERGO CAN PLAY A ROLE)

In the world of ergonomics and safety, we are well versed in leading and lagging indicators. According to the U.S. Occupational Safety and Health Administration,

DOI: 10.1201/9781032658988-19

leading indicators are proactive, preventive, and predictive measures that provide insight into the effectiveness of safety/ergonomics/health activities. They track events preceding incidents, such as injuries or illnesses. When identified as drivers, tracking them helps to identify potential problems within a program. Unlike lagging indicators, which measure past occurrences, leading indicators enable organizations to take preventive actions before issues escalate into incidents.

To identify effective leading key performance indicators that visualize progress and assess the achievement of main goals, it is crucial to first determine the actions or activities that influence and drive these indicators. Just as in ergonomics, it is essential to know what to track to drive actions and what to measure for accountability when identifying metrics for DEI.

Considering the guidelines and best practices for selecting and tracking DEI-related metrics, how can ergonomics, or more specifically, which ergonomics-related metrics could be used to measure an organization's success in achieving its DEI goals?

For starters, the tremendous impact that our ergonomics efforts have on health and safety objectives (elimination or minimization of injury occurrences and severity; high-level maturity safety culture; implementation of total worker health; improved employee (and family) biometrics, etc.) is one area we should track. Why? Because those safety objectives, including participatory ergonomics efforts, have shown correlations to business outcomes, e.g., improved employee retention (Aman-Ullah et al., 2022; Lallemand, 2012), improved employee well-being, reduced absence and presenteeism (Rolls Royce, Abraham et al., 2023; Machado, 2023), and employee engagement (Biddison et al., 2016). Additionally, there is the direct impact ergonomics efforts have on operational productivity and efficiency (Vink & Molenbroek, 2006; Brunner & Bengler, 2023). All areas tie back to the DEI metrics recommended in the list of best practices. But of course, the more data we can provide with those metrics to drive actions, such as the identification of any disparity gaps with regard to group demographics (e.g., gender) and the relation between high-risk jobs and employee retention, the better. This data-driven information (particularly descriptive statistics and relevant correlations) could identify prioritization opportunities that could better guide decisions about the allocation of resources (for ergonomics process, DEI, operations, etc.) and provide quantitative means to track progress of targeted DEI initiatives.

Additionally, recognizing and tracking metrics tied to financial business outcomes goes a long way with showing impact and value. Talk about "COGS" and "Gross Profit Margin" and the like is not normally in our lexicon, but it should be. While it is important to continue to "humanize" safety and ergonomics communications and keep the worker at the forefront of our metric conversations (i.e., when talking about safety metrics speak in terms of the person as opposed to just the rate or cases (e.g., "the plant's TCIR) rate was 2.5, meaning 2.5 out of a 100 full time employees had a recordable injury"), it is still important to know and understand where our efforts can have an impact on other areas of the organization. That means tracking financial impacts of ergonomics and DEI efforts in terms of direct and indirect costs associated with injury reduction (e.g., reduction in workers compensation costs (be mindful to assign these costs as "cost avoidance") or any insurance rate premium reduction as a result of improved safety performance), productivity (e.g., work standard time savings), quality (e.g., defect reductions), procurement cost reductions (e.g., mitigation lead to the elimination of required tools or PPE), or innovation (e.g., return on investment of frugal ergonomics solutions or revenue from any new business venture as a result of participatory engagement in innovation initiatives).

Of course, there are opportunities to move the needle on other business outcomes, such as with sustainability (e.g., collaborative DEI & ergonomics projects can be highlighted in sustainability reports and captured in ESG reporting) and marketing (e.g., the visibility of ergonomics projects at global conferences such as the Applied Ergonomics Conference Ergo Cup has the potential to positively impact an organizations branding/public image and can also lead to positive brand sentiment). See Table 15.1 for how metrics captured as part of the ergonomics process could be beneficial for DEI strategies — all great opportunities to highlight the potential value we can bring to an organization!

KEEPING DEI PROGRESS ON ERGO'S RADAR

As discussed in previous sections of this book, DEI principles are already deeply embedded in the core of ergonomics practices. A robust ergonomics process routinely designs solutions that cater to diverse needs, manage risks, and promote inclusion, aligning closely with DEI objectives. However, ergonomics practitioners can do more in this aspect, such as working closely with departments such as Human Resources, Safety, and Operations to ensure that ergonomics solutions not only address physical and psychosocial safety but also support the organization's broader DEI commitments.

Therefore, it will be critical for us, as ergonomics practitioners and as a profession, to stay up to date and abreast of all organizational DEI strategies and their progress, as well as broader DEI-related standards and regulations. This ongoing awareness ensures that ergonomics practices are aligned with evolving DEI goals and compliance requirements. For example, in the case of managing a program that encompasses any form of functional testing, it is imperative that we have a thorough understanding of local laws and regulations such as the United States' EEOC Uniform Guidelines or Americans with Disabilities Act, as that can impact the legality and effectiveness of said functional testing (whether post-offer or pre-hire testing). Additionally, keeping up to date with your organization's DEI strategies allows us, as ergonomics practitioners, to further and more effectively contribute to company-wide initiatives. For example, if an organization adopts a new DEI policy emphasizing the importance of creating inclusive environments for neurodivergent employees, ergonomics practitioners can ensure that ergonomics design solutions reflect and can effectively demonstrate those priorities.

Staying keyed into broader DEI-related standards and best practices (such as ISO 30415 as discussed in Chapter 5) can inspire us, as ergonomics practitioners, to incorporate more innovative approaches into our initiatives that further enhance inclusion. One way to stay connected is to incorporate DEI-related metrics into our

TABLE 15.1
Ergonomics Specific Metrics That Could be Beneficial to Monitor as Part of DEI Strategies

Ergo-Related Metric	Description	DEI Benefit / Opportunity	Key DEI Theme(s)	Ergo Benefit / Opportunity
Total Recordable Incident Rate (TRIR)	aka "OSHA Rate"; calculation of ((total recordable cases × 200,000) / total hours worked) × 100	Identify potential disparities with certain groups (as applicable by: location, department, job/task, tenure, seniority, gender, age, etc..)	Recruitment. Retention, Compliance	Identify potential areas of disparities to consider in priority setting
Days Away, Restriction, or Transferred (DART) IR	cases resulting in Days Away, Restriction, or Transferred; calculation of ((total DART cases × 200,000) / total hours worked) × 100	(Same benefit / opportunity)	Recruitment. Retention, Compliance	(Same benefit / opportunity)
Serious Injury or Fatality (SIF) IR	Serious Injury or Fatality ; calculation of ((total SIF cases × 200,000) / total hours worked) × 100	(Same benefit / opportunity)	Recruitment. Retention, Compliance	(Same benefit / opportunity)
MSDs IR	((total MSD recordable cases × 200,000) / total hours worked) × 100	Identify potential disparities with certain groups specifically related to work/ work place design (as applicable by: location, department, job/task, tenure, seniority, gender, age, etc.)	Inclusion, Recruitment. Retention	Identify potential areas of disparities to consider in priority setting; identify opportunities for inclusive design improvement collaboration projects
MSDs DART IR	((total MSD DART cases × 200,000) / total hours worked) × 100	(Same benefit / opportunity)	Inclusion, Recruitment. Retention	(Same benefit / opportunity)
MSDs SIF IR	((total MSD SIF cases × 200,000) / total hours worked) × 100	(Same benefit / opportunity)	Inclusion, Recruitment. Retention	(Same benefit / opportunity)

(Continued)

TABLE 15.1 (CONTINUED)
Ergonomics Specific Metrics That Could be Beneficial to Monitor as Part of DEI Strategies

Ergo-Related Metric	Description	DEI Benefit / Opportunity	Key DEI Theme(s)	Ergo Benefit / Opportunity
Median Workers Comp Case Cost per MSD case	Median Workers Comp Case Cost per MSD case	(Same benefit / opportunity)	Inclusion, Recruitment. Retention	(Same benefit / opportunity)
(If applicable) % Insurance Premium Rate Change (annual)	Annual comparison of insurance premium rate change	(Same benefit / opportunity)	Development & Advancement; Market Impact	(Same benefit / opportunity)
% of High Risk Tasks	Percentage of total job tasks assessed as HIGH RISK via selected ergonomics evaluation process	(Same benefit / opportunity)	Inclusion, Recruitment. Retention, Compliance	(Same benefit / opportunity)
% Population Accommodated by Task	Percentage of (local) male/female population accommodated in the task	(Same benefit / opportunity)	Inclusion, Recruitment. Retention, Development & Advancement	(Same benefit / opportunity)
% of Standard Jobs with Completed / Updated Risk Assessments	Percentage of total standard job tasks assessed via selected ergonomics evaluation process	Identify potential disparities with equitable risk management (as applicable by: location, department, job/task, tenure, seniority, gender, age, etc.)	Inclusion, Engagement, Retention	Identify potential disparities with equitable risk management; identify potential areas of disparities to consider in project planning; Identify potential areas of disparities to consider in priority setting; identify opportunities for inclusive design improvement collaboration projects

(*Continued*)

TABLE 15.1 (CONTINUED)

Ergonomics Specific Metrics That Could be Beneficial to Monitor as Part of DEI Strategies

Ergo-Related Metric	Description	DEI Benefit / Opportunity	Key DEI Theme(s)	Ergo Benefit / Opportunity
% of Non-Standard Tasks Identified with a Completed / Updated Risk Assessments	Percentage of total non-standard tasks identified and assessed via selected ergonomics evaluation process	(Same benefit / opportunity)	Inclusion, Engagement, Retention	(Same benefit / opportunity)
# of High Risk Task Reductions	Count of improvements significant to reduce risk level; assessment of effectiveness of ergonomics efforts	(Same benefit / opportunity)	Inclusion, Productivity, Retention	(Same benefit / opportunity)
# of Ergonomics Improvements	Count of ergonomics improvements, including improvements not significant enough to reduce risk level or an improvement or elimination to a MODERATE or LOW level risk; assessment of short term	(Same benefit / opportunity)	Inclusion, Productivity, Retention	(Same benefit / opportunity)
% of Sites/ Departments with Ergo Team a/o Kaizen Plan	Percentage of locations able to allocate resources to manage ergonomics efforts	(Same benefit / opportunity)	Inclusion, Engagement, Retention	(Same benefit / opportunity)
% of solutions by hazard control category (elimination, engineering, admin, PPE)	Percentage of solutions by hazard control category	(Same benefit / opportunity)	Productivity	(Same benefit / opportunity)

(Continued)

TABLE 15.1 (CONTINUED)
Ergonomics Specific Metrics That Could be Beneficial to Monitor as Part of DEI Strategies

Ergo-Related Metric	Description	DEI Benefit / Opportunity	Key DEI Theme(s)	Ergo Benefit / Opportunity
% of High Risk Tasks with Identified Solution	Percentage of tasks assessed as HIGH RISK via selected ergonomics evaluation process with an identified solution (awaiting resources for implementation)	(Same benefit / opportunity)	Inclusion, Productivity, Engagement	(Same benefit / opportunity)
Average Time to Close Planned Action Items	Average time taken to close activities planned for the time period	(Same benefit / opportunity)	Productivity, Retention	(Same benefit / opportunity)
% of Sites/ Departments with Prioritized Top 5	Percentage of locations able to identify priority for ergonomics efforts	(Same benefit / opportunity)	Inclusion, Engagement	(Same benefit / opportunity)
% of Sites/ Departments with Action Plan for Prioritized Top 5	Percentage of locations able to allocate resources to manage ergonomics efforts on prioritized risks	(Same benefit / opportunity)	Inclusion, Engagement, Retention	(Same benefit / opportunity)
% Employees Trained in Ergonomics (Basic)	Percentage of employees trained in (basic) ergonomics	(Same benefit / opportunity)	Productivity, Engagement, Retention	(Same benefit / opportunity)
% Management Trained in Ergonomics	Percentage of managers trained in ergonomics specifically targeting management	(Same benefit / opportunity)	Productivity, Engagement, Retention	(Same benefit / opportunity)
% Engineering Trained in Ergonomics	Percentage of engineers trained in ergonomics specifically targeting engineering/designers	(Same benefit / opportunity)	Productivity, Engagement, Retention	(Same benefit / opportunity)

(Continued)

TABLE 15.1 (CONTINUED)
Ergonomics Specific Metrics That Could be Beneficial to Monitor as Part of DEI Strategies

Ergo-Related Metric	Description	DEI Benefit / Opportunity	Key DEI Theme(s)	Ergo Benefit / Opportunity
% of employees rating site/ department at × level of safety maturity	Percentage of employees rating site/ department at or above × level of safety maturity level (e.g., Bradley Curve)	(Same benefit / opportunity)	Inclusion, Engagement	(Same benefit / opportunity)
# New Processes or Machines Installed at Low Risk	Count of new installations (process, machine, equipment, etc.) assessed as LOW RISK via selected ergonomics evaluation process or design change management	Assessment of inclusive design; identify potential disparities with equitable risk management	Inclusion, Productivity, Retention	Identify potential areas of disparities to consider in priority setting
% of New Processes or Machines Installed With Ergo Team Review	Percentage of new installations (process, machine, equipment, etc.) with employee review	Assessment of employee engagement with identifying solutions specific to local workforce needs for inclusive design; assessment of innovation culture (as applicable by: location, department, job/task, tenure, seniority, gender, age, etc.)	Inclusion, Productivity, Engagement, Retention	Identify potential disparities with equitable risk management; identify potential areas of disparities to consider in project planning; Identify potential areas of disparities to consider in priority setting; identify opportunities for inclusive design improvement collaboration projects

(Continued)

TABLE 15.1 (CONTINUED)
Ergonomics Specific Metrics That Could be Beneficial to Monitor as Part of DEI Strategies

Ergo-Related Metric	Description	DEI Benefit / Opportunity	Key DEI Theme(s)	Ergo Benefit / Opportunity
# of solution development events (e.g., kaizens)	Count of planned cross-functional solution development events (with employee participation)	(Same benefit / opportunity)	Inclusion, Productivity, Engagement, Retention, Development & Advancement	(Same benefit / opportunity)
# of identified solutions generated in solution development events	Count of identified solutions in planned cross-functional solution development events	(Same benefit / opportunity)	Inclusion, Productivity, Engagement, Retention	(Same benefit / opportunity)
Average ROI ($) of identified solutions generated in solution development events	Average ROI of identified solutions in planned cross-functional solution development events	(Same benefit / opportunity)	Inclusion, Productivity, Engagement, Development & Advancement	(Same benefit / opportunity)
# of patentable ergonomics solutions	Count of ideas developed through the ergonomics process that can be considered for patenting (via organization's IP process)	(Same benefit / opportunity)	Inclusion, Productivity, Engagement, Retention, Development & Advancement	(Same benefit / opportunity)

(Continued)

TABLE 15.1 (CONTINUED)
Ergonomics Specific Metrics That Could be Beneficial to Monitor as Part of DEI Strategies

Ergo-Related Metric	Description	DEI Benefit / Opportunity	Key DEI Theme(s)	Ergo Benefit / Opportunity
% of employees participating in innovation initiatives	Percentage of employees participating (submitting ideas, voting on ideas, etc.) in innovation initiatives / targeted campaigns	(Same benefit / opportunity)	Inclusion, Engagement, Development & Advancement	(Same benefit / opportunity)
% of solutions with specific DEI benefit(s) identified	Percentage of ergonomics solutions specifically delineating DEI benefit(s) (e.g., knurling added improving accessibility for workers with arthritis upper extremity	Details about inclusive design; opportunity to share as part of organization sustainability report	Inclusion, Recruitment, Retention	Opportunity to show value of ergonomics solutions in general and inclusive design, particularly solutions made more accessible (impact to all); identify opportunities for inclusive design improvement collaboration projects
Average % Efficiency Gain per implemented solution	Average percentage productivity gain realized from implemented solutions	Opportunity to show value of inclusive design	Productivity, Market Impact	Opportunity to show value of ergonomics solutions in general and inclusive design, particularly solutions made more accessible (impact to all)

(*Continued*)

TABLE 15.1 (CONTINUED)
Ergonomics Specific Metrics That Could be Beneficial to Monitor as Part of DEI Strategies

Ergo-Related Metric	Description	DEI Benefit / Opportunity	Key DEI Theme(s)	Ergo Benefit / Opportunity
Average % Defect Reduction per implemented solution	Average percentage quality gain realized from implemented solutions	(Same benefit / opportunity)	Productivity, Market Impact	(Same benefit / opportunity)
% Accommodation Improvement With Implemented Solutions	Percentage of increase in of (local) male/female population accommodated in the task with implemented solutions	(Same benefit / opportunity)	Inclusion, Recruitment, Retention, Compliance, Development & Advancement	(Same benefit / opportunity)
Average % Improvement in Communications Positive Sentiment Score (for events/communications related to ergonomics initiatives (e.g., AEC Ergo Cup))	Average percentage improvement in communications Positive Sentiment Score for any branding event associated with ergonomics	(Same benefit / opportunity)	Market Impact, Development & Advancement	(Same benefit / opportunity)
# of employee-originated solutions submitted	Count of employee-originated solutions submitted	Assessment of employee engagement	Inclusion, Engagement, Retention	Identify potential areas of disparities to consider in project planning (e.g., kaizen planning); identify potential areas of disparities to consider in priority setting

(*Continued*)

TABLE 15.1 (CONTINUED)
Ergonomics Specific Metrics That Could be Beneficial to Monitor as Part of DEI Strategies

Ergo-Related Metric	Description	DEI Benefit / Opportunity	Key DEI Theme(s)	Ergo Benefit / Opportunity
# of employee-originated solutions implemented	Count of employee-originated solutions implemented	(Same benefit / opportunity)	Inclusion, Engagement, Retention	(Same benefit / opportunity)
# of interdisciplinary collaboration projects	Count of interdisciplinary collaboration projects (e.g., ergonomics team/practitioner consulting Disability BRG regarding the design of a new office to design for safety, efficiency and inclusiveness (e.g., inclusion of adjustable desks and quiet spaces for neurodiverse employees))	Assessment of employee engagement; details about inclusive design; opportunity to share as part of organization sustainability report	Inclusion, Engagement, Retention	Identify potential areas of disparities to consider in project planning (e.g., collaboration projects with ERGs)
eNPS / Survey Score on Workplace Design Satisfaction	Survey score or eNPS (employee Net Promoter Score) assessing satisfaction on workplace design	Assessment of sense of inclusion; opportunity of identify potential disparities with certain groups specifically related to work/work place design (as applicable by: location, department, job/task, tenure, seniority, gender, age, etc.)	Inclusion, Engagement, Retention	Identify potential areas of disparities to consider in priority setting; identify potential areas of disparities to consider in project planning (e.g., training plans)
eNPS / Survey Score on Accommodation/RTW process	Survey score or eNPS (employee Net Promoter Score) assessing satisfaction of the accommodation process	(Same benefit / opportunity)	Inclusion, Engagement, Retention	(Same benefit / opportunity)

(Continued)

TABLE 15.1 (CONTINUED)
Ergonomics Specific Metrics That Could be Beneficial to Monitor as Part of DEI Strategies

Ergo-Related Metric	Description	DEI Benefit / Opportunity	Key DEI Theme(s)	Ergo Benefit / Opportunity
% of Accommodations / RTW with Ergonomics Review	Percentage of accommodation cases or return to work cases with ergonomics engagement expertise	Assessment of sense of inclusion; details about inclusive design; opportunity to share as part of organization sustainability report	Inclusion, Engagement, Retention	Identify potential areas of disparities to consider in ergonomics process resource planning
(If applicable) Fail rate per functional test	Rate of failure for each function testing protocol	Assessment of sense of inclusion; details about inclusive design	Inclusion, Engagement, Recruitment	(Same benefit / opportunity)
Count of Other Domain Initiatives with Ergonomics Review	Count of Other Domains (e.g., HR/ Employee Relations, Compensation, Medical, etc.) initiatives (e.g., shift work scheduling, functional testing, feminization) with ergonomics engagement expertise	(Same benefit / opportunity)	Inclusion, Engagement, Retention	(Same benefit / opportunity)

overall ergonomics processes. These metrics could help ergonomics practitioners to monitor and evaluate the impact of ergonomics practices on the diversity, equity, and inclusion (separately) of their workforces, as well as to identify areas for improvement and innovation. For instance, we may be able to leverage data from DEI-related metrics to identify areas where ergonomics solutions can be improved to better support underrepresented groups—such as the feminization efforts undergoing in the agriculture sector in Brazil (Brazilian Farmers, 2023) and Mexico (Kovoleva, 2021) to increase the number of women in these traditionally male-dominated trades. This could include using ergonomics assessments to address specific challenges faced by women in those jobs, thus aligning ergonomics efforts with the organization's commitment to equity across all levels.

However, as a caution, Jourdan finds that it can be challenging to measure certain aspects of DEI from a data perspective because a lot of the data is qualitative, especially when it comes to "feelings of inclusion." However, she notes it is important for companies to figure out what metrics are relevant and reportable that can show the changes or obstacles that the workplace faces. She shares,

> ...often the discussions around DEI are based on lived experiences, which are very significant and valuable, but there is not enough data to support them, like what percentage of the workforce belongs to a certain group, or what kind of data is available, but sometimes the sample size is too small.

So, how are DEI-related metrics qualitative and/or quantitative enough to be able to effectively track progress and effectiveness yet adequately assess those lived experiences captured and tracked? More importantly, how can ergonomics help? For starters, tapping into the voice of the employees through engagement surveys and the ERGs themselves would be valuable sources of data and feedback that could be parsed to create an aggregate of sentiments. These surveys and targeted feedback channels with the ERGs could be tailored to measure how employees feel about their work environment, their sense of belonging, their perception of fairness and respect, and their awareness and participation in DEI initiatives.

Beyond helping to champion the collection of data with and through our participatory approach, ergonomics practitioners can definitely play a role in crafting questions by applying principles of cognitive ergonomics and human factors. For example, we could provide guidance on how to design questions that are clear, concise, and engaging for the respondents while minimizing ambiguity and bias in the questions, to ensure that they are culturally and linguistically appropriate for the target audience, all in a way that facilitates frank responses.

We could also aid in the communication of results, ensuring that information is shared in ways that can better reach a diverse workforce—such as providing guidance on inclusive language, varied mediums, and effective utilization of visual and audible aids to aid in accessibility (see Part 2 for more information on accessible communications).

No matter what kind of quantitative metrics are selected to measure and communicate DEI strategy progress, they should not be the only source of data, but rather one part of a mixed approach that also includes qualitative data and background information that can give a more comprehensive and detailed insight into the complexities and nuances of DEI issues.

But what are other non-traditional metrics that would be beneficial for us, as ergonomics practitioners and advocates for DEI? Typically speaking, the metrics previously discussed are often captured in standard dashboards for corporate DEI strategies, with the inclusion of other lagging indicators grounded in the organization's values and mission that DEI efforts impact. But if we have the chance to dig deeper as ergonomics practitioners to truly gauge our ergonomics efforts, what could be the metrics we track or at least be aware of?

For starters, it would be beneficial to understand demographic data as related to specific tasks and positions that may be physically and/or mentally taxing in a way that disproportional impacts or systematically excludes groups of workers—such as jobs and tasks that are gender dominated due to certain physical demands, or jobs with low levels of autonomy segregated by socioeconomic status due to lack of desirability. The visibility of how work affects different groups, and its subsequent impact on recruitment and retention, particularly attrition or turnover data, can be factored into ergonomics priority setting and resource planning. And potentially that level of information could prove to be useful for more detailed risk analysis and mitigation. Beyond that, that level of detail could be expanded, in the form of more robust physical demands analyses (PDAs) and functional job descriptions. By doing that, organizations can identify roles and positions that could be made more inclusive, opening opportunities to a wider pool of candidates. For example, positions that traditionally require specific cognitive abilities could be adjusted to accommodate individuals with cognitive disabilities, ensuring that essential functions are not unnecessarily restrictive, such as rethinking the way tasks are performed, offering alternative methods or assistive technologies that support cognitive processing.

We should also dig deeper into collecting and understanding lived experiences, especially as they relate to the work we design and mitigate risk for. This is because it is important to reflect on the diversity of work itself. It is one thing to assess and mitigate risk of work based on expectations of "work as designed", but we also must strive to understand how "work is actually done" and assess/mitigate to reality, including deviations from the norm which can be significantly impacted by cultural and physical diversity. In speaking with French ergonomist Didier Lafont he expresses a great point as related to DEI (translated from French):

> Diversity, Equity, and Inclusion are not static states that we achieve after a process, but rather they are a construction, that is constantly challenged by the changes in the organization and work.

To me, that speaks to why we, as ergonomics practitioners, constantly need to keep a pulse of our respective workforce's expectations and perceptions of work (and risk), and that includes incorporating diversity of not only the workforce tasked to do and respond to work, but the ever-changing diversity of work itself due to technological, organizational, or environmental factors. That involves collaborating with DEI initiatives and DEI-related metrics to obtain and track feedback on how workers perceive work, their expectations, the norms they operate within (and how they diverge from the standard, or how work was "designed") and use that information to continuously improve. By doing so, we can ensure that our ergonomics solutions are effective and respectful of the different needs and situations of our workers and the work they do. Also, by understanding how work is done by different cultures we can identify and address potential sources of bias that may create barriers or challenges for some workers, such as stereotypes, prejudices, microaggressions, or implicit assumptions that could affect their performance and well-being.

With all the potential they have for improving our ergonomics processes and giving us ways to show more value, DEI metrics are essential for our ergonomics efforts. By embedding the measurement and reporting these metrics into our ergonomics processes, we would be building sustainable systems that demonstrate our commitment and contribution to the organizational goals and values related to DEI. Below are DEI-related metrics that should be on your ergonomics process radar (if you're not already tracking):

- Recruitment Metrics—especially:
  - Hiring trends by department, especially as related to challenge jobs or in connection with high-risk jobs (to validate priority levels).
- Retention Metrics—especially:
  - Voluntary Attrition/turnover rates by job, gender, other demographic, especially as related to high demand or high-risk jobs (to validate priority levels).
  - Representation within physically demanding jobs (e.g., gender, age, physical and cognitive abilities, racial, socioeconomic)
- Employee Resource Groups Engagement Metrics: As related to the ERG's engagement in workplace design decisions
- Accommodations Metrics: As related to types of accommodations, effectiveness of solution, satisfaction survey from employees, etc.—specifically accommodations that involved with the ergonomics process; types of accommodations, effectiveness of solution
- Employee Experience Metrics: Employee Net Promoter Score (eNPS) or satisfaction survey score (by department, work group, etc.) as related to:
  - Satisfaction with the overall employee experience
  - Satisfaction with workplace design
  - Satisfaction with well-being
  - Satisfaction with the job satisfaction
- Disability Index (specifically):
  - Culture and Leadership: Assessing the commitment of leadership to disability inclusion and the presence of disability-focused ERGs. Opportunity for the ergonomics process to identify potential collaboration efforts re: design, risk abatement.
  - Enterprise-Wide Access: Evaluating the accessibility of facilities, technology, and resources for employees with disabilities. Opportunity for

the ergonomics process to more effectively prioritize efforts (e.g., focus on broadening the candidate pool through improving accessibility).
    - Employment Practices: Reviewing recruitment, hiring, and retention practices to ensure they are inclusive of people with disabilities. Opportunity for the ergonomics process to more effectively prioritize efforts (e.g., identify roles that may need to be an ergonomics priority).
    - Community Engagement: Measuring the company's involvement in disability-related community initiatives and partnerships. Opportunity for the ergonomics process to more effectively prioritize and plan program efforts (e.g., extend communications about ergonomics through providing expertise—volunteer opportunities).
- Business Impact: Likelihood to recommend.
- Business Impact: Customer satisfaction.
- Business Impact: ROI product/service development.

Of course, the list of metrics shared is not exhaustive, nor will all be applicable to your site or organization. However, adopting a broader perspective would be valuable for both strategies and will better solidify the goal of developing a cohesive organizational culture of inclusiveness. Tracking many of these metrics often wouldn't add extra work for the ergonomics process, as they are typically captured as part of "doing ergonomics business."

Reviewing these metrics at the very least presents an opportunity for ergonomics practitioners to review their program management and possibly identify metrics that should be included or amended to better assess their program's efforts in contributing to DEI and, by extension, their organizational strategic goals. This is another way to demonstrate our value and build "stickiness" (i.e., keep them coming back to the ergonomics well).

USING ERGONOMICS/HUMAN FACTORS PRINCIPLES TO DEVELOP DEI METRICS

Beyond the potential to incorporate ergonomics metrics within the overall scope of measuring the success of DEI (through all the collective initiatives designed to positively impact DEI, including improved application of ergonomics in the workplace), there is also the potential to deploy ergonomics principles in the development of said DEI metrics such that they are tailored to the organization through a user-centered approach. Bruni et al. (2023), citing the lack of a standardized corporate DEI metric, has proposed a five-step process (that consists of research, brainstorming, synthesizing, generation, and deployment) for organizations to develop DEI metrics best suited to their specific needs, goals and internal culture.

The steps in the proposed process rely heavily on an understanding of the current state of the organization and their goals for the future state, as it is these gaps that determine ultimately what specific metrics should be captured as part of the glide path for improvement. An additional aspect of this process is the level of engagement required. By getting the participation of stakeholders, such as members of the DEI team ("IDEA" in the case study), it allows for direct engagement of those tasked to tap into the realities of the work culture ("work as done versus work as perceived"). This collaborative, user-centered approach ensures that the development of DEI metrics is not only aligned with organizational needs but is also rooted in the lived experiences of employees, allowing for more effective and actionable outcomes.

POTENTIAL NEGATIVE ATTITUDES TOWARD DEI METRICS

While metrics are important to assess and consequently improve progress for DEI efforts, any metric around DEI is not without controversy, mainly due to misconceptions and misinformation of how and why data is collected. One misunderstanding that often comes up with DEI metrics is the idea of "quotas" (Okeh, 2022) or a kind of tokenism or favoritism (Gaines, 2024), which could imply that organizations value demographic or identity representation more than merit or qualifications in hiring, promotion, or decision-making processes (Okeh, 2022)—the implication that viewpoint has that certain demographics lack said merit or qualifications notwithstanding, which is an unfair assumption that overlooks the systemic challenges that have historically limited opportunities for these groups. Nevertheless, some people have the added view that DEI efforts are an imposition of certain ideology. Such viewpoints, unfortunately held by many as can be seen with anti-DEI movements across the globe, could damage the trust and morale of the workforce and cause resentment or backlashes among those who feel left out or discriminated against because of it.

These challenges can cause people to resist or oppose anything related to DEI, or worse, undermine or sabotage the efforts, and slow down the progress. From my perspective, I feel the opportunity lies in effective communication. Those that may think that DEI is only about "meeting quotas" based on demographic characteristics (as opposed to creating more fair opportunities and addressing systemic barriers to inclusion) make DEI harder to communicate and get support for, not to mention motivate action. However, we need to meet people where they are if we want to make progress in this area. We need to understand that many people have real fears that need to be acknowledged. This does not mean we have to stop DEI progress, instead, it means we must adapt to reduce the risk, either at the source (fear) or the symptom (pushback or complacency).

We need to focus on clear and transparent communication that explains the purpose, strategy, and expected outcomes of our DEI efforts regarding the data collection and analysis used to measure DEI progress, especially as related to DEI data captured for ergonomics (and safety) when it may not have been captured (or communicated in such a way) before. We also need to focus on what metrics should be measured and how the data is collected to ensure an accurate and true representation of the workplace and workforce situation. Most importantly, we need to use a participatory approach, like the approach with ergonomics programs (the design of which is adapted to the specific needs and resources available) to ensure the voice of the workers is diverse and drives all DEI and ergonomics efforts, including progress management.

REFERENCES

Abraham, L., Russell, R., Sanchez-Riera, L., Emir, B., Roomes, D., Green, K., Taylor-Stokes, G., Mirams, L., Wallis, H., & Burton, K. (2023). Quantifying the burden of persistent musculoskeletal pain in employees at Rolls-Royce in the United Kingdom: A real-world cross-sectional survey. *Journal of Occupational and Environmental Medicine*, *65*(6), e424—e434. https://doi.org/10.1097/JOM.0000000000002845

Aman-Ullah, A., Ibrahim, H., Aziz, A., & Mehmood, W. (2022). Impact of workplace safety on employee retention using sequential mediation: Evidence from the health-care sector. *RAUSP Management Journal*, *57*(2), 182–198.

Brazilian Farmers. (2023, June 21). *For more women in agribusiness and foreign trade.* Brazilian Farmers. https://brazilianfarmers.com/news/for-more-women-in-agribusiness-and-foreign-trade/

Biddison, E. L. D., Paine, L., Murakami, P., Herzke, C., & Weaver, S. J. (2016). Associations between safety culture and employee engagement over time: A retrospective analysis. *BMJ Quality & Safety*, *25*(1), 31–37.

Bruni, S., Shenberger-Trujillo, J., Clark, T., Erickson, I., Toumbeva, T., & Meyer, S. (2023, September). Developing a human-centered corporate metric for inclusion, diversity, equity, & antiracism in a small business: A case study and lessons learned. In *Proceedings of the Human Factors and Ergonomics Society Annual Meeting* (Vol. 67, No. 1, pp. 1451–1456). SAGE Publications. https://doi.org/10.1177/21695067231192523

Brunner, S., Kühnel, K., & Bengler, K. (2023). Lean Ergonomics—an empirical combination of Management Science and Ergonomics. *Zeitschrift für Arbeitswissenschaft*, *77*(4), 515–526.

Coutelier, L., & Engfeldt, H. (2022, May 22). *Running a privacy-law-compliant inclusion and diversity data collection program globally.* International Association of Privacy Professionals. https://iapp.org/news/a/running-a-privacy-law-compliant-inclusion-and-diversity-data-collection-program-globally/

Gaines, K. (2024, August 26). *From tokenism to true inclusion: The essential role of equity in DEI.* Forbes. https://www.forbes.com/councils/forbesbusinesscouncil/2024/08/26/from-tokenism-to-true-inclusion-the-essential-role-of-equity-in-dei/

Kovaleva, M. (2021). Feminization of agriculture and its impact on sustainable development. In W. Leal Filho, A. Marisa Azul, L. Brandli, A. Lange Salvia, & T. Wall (Eds.), *Gender equality* (pp. 901–912). Encyclopedia of the UN Sustainable Development Goals. Springer. https://doi.org/10.1007/978-3-319-95687-9_92

Lallemand, C. (2012). Contributions of participatory ergonomics to the improvement of safety culture in an industrial context. *Work*, *41*(Supplement 1), 3284–3290.

Machado, M. (2023). Impact and benefits of ergonomic interventions at the workplace. *Journal of Ergonomics*, *13*, 354.

Okeh, N. (2022, September 27). *Diversity quotas face challenges and legal risk.* Practical ESG. https://practicalesg.com/2022/09/diversity-quotas-face-challenges-and-legal-risk/

Thompson, S. (2017, June 30). *Defining and measuring 'inclusion' within an organisation.* K4D Helpdesk. https://www.researchgate.net/publication/343306308_Defining_and_measuring_'inclusion'_within_an_organisation

Vink, P., Koningsveld, E. A. P., & Molenbroek, J. F. (2006). Positive outcomes of participatory ergonomics in terms of greater comfort and higher productivity. *Applied Ergonomics*, *37*(4), 537—546. https://doi.org/10.1016/j.apergo.2006.04.012

Closing

If you have found your way to the end of this book, I will have to assume you *now* have at least some level of interest in ergonomics and Diversity, Equity, and Inclusion (DEI). And hopefully, you see a clear connection between the two domains and how both can be used to benefit each other. It just makes good business sense!

This is especially true when you consider the impact having an engaged workforce has on an organization—its productivity and its long-term sustainability in a competitive world. And creating an inclusive workplace that supports employee needs is essential for fostering this engagement. In my conversation with Anthony Banks, he highlighted the key role ergonomics plays in a company's success through the creation of inclusive workspaces. By aligning work environments with human capabilities and their needs for both physical and psychological safety, ergonomics can reduce injury risks (thus turnover) and improve productivity while enhancing employee well-being. And in a competitive job market, he added, where top talent has many options, employee well-being will be a key factor in attracting and retaining skilled workers.

> In today's shifting landscape, top job seekers can choose from multiple companies and opportunities.

This is supported by studies such as Lööw's research on ergonomics and workplace attractiveness (Lööw, 2019) and Hoffmeister et al.'s findings on the link between ergonomics and talent retention (Hoffmeister et al., 2015). These studies show that organizations prioritizing ergonomics and wellness signal that they value their workers as individuals and are more likely to retain that talent.

So, as Anthony argued, prioritizing ergonomics is a strategic move, especially when it is connected to the benefits for DEI as that boosts workforce health and enhances organizational competitiveness. And as workspaces and demographics continue to evolve, addressing these new variables presents new challenges. And meeting diverse needs, on top of continuing to improve safety and operational efficiency, will become a necessity for organizational sustainability. However, this shift will also afford ergonomics practitioners the opportunity to demonstrate the value we bring as problem solvers who will help organizations create inclusive environments and foster the innovation needed to address these new challenges to stay viable in a competitive market. Like I said, it just makes good business sense!

CALLS TO ACTION

While this may be the conclusion to this book, I hope that it is not the conclusion of your journey of advancing DEI efforts, whether it is in your organization or in

DOI: 10.1201/9781032658988-20

another capacity. And hopefully, this book has given you the confidence to tackle the challenges ahead in this journey—be it to further embed DEI efforts within your organization's ergonomics process, identifying technical solutions to meet a specific employee need or initiatives to accommodate a widely diverse workforce, or identifying ways to influence others within and around your organization to continue to fight through "anti-DEI" noise to recognize the vast benefits for all that can be acheived. To help you on that journey, I have a few calls to action:

Calls to Action for the Ergonomics Profession

Marketing, Marketing, Marketing!

Historically, we have done a commendable job of aligning ourselves with safety and health concerns. In fact, we have been so good at that that we are seen as only the de facto "protector of employees" (Mulaomerovic, 2023). As authors Dul et al. pointed out, ergonomics *must* demonstrate its value more effectively to stakeholders who have a significant influence on system design, including key decision-makers (Dul et al., 2012). So, it is high time we kicked our public relations acumen into high gear to show our value, which includes being a "can't function without" strategic contributor to business outcomes.

To do this, it becomes imperative that we expand our reach beyond our traditional "ergo" circles. Our conferences are fantastic (and we should absolutely continue holding, participating in, and inviting others to them, as they do a great job of keeping our "ergo" circles up to date!). However, we need to do more to educate other industries and professions. Just as we have made significant inroads in health and safety, we should aim to have a strong presence in Human Resource, Legal, and other relevant sectors. In fact, as a good barometer, wherever DEI goes, ergonomics should follow as the technical arm that can help organizations meet their DEI goals.

Get Engaged, Stay Engaged

We should also support our professional societies in their DEI outreach efforts. In conversations with Brian Sherman and Shandrea Middleton, co-chairs of the DEI sub-committee within the Applied Ergonomics Society (AES), they shared their dedication to making ergonomics more inclusive. Through initiatives like reaching out to Historically Black Colleges and Universities, the AES seeks to engage students at both the graduate and undergraduate levels. Additionally, collaborations between the Human Factors and Ergonomics Society (HFES) and the Chartered Institute of Ergonomics and Human Factors aim to provide further guidance on how ergonomics can address DEI issues (Wooldridge et al., 2022). It is important for us all to participate, whether through member surveys, outreach, or speaking on DEI panels (or taking that panel on the road to one of the HR conferences—wink wink!), to contribute wherever we can.

Calls to Action: Ergonomics Practitioners (or Advocates)

Prioritize Having an Inclusive Design Mindset

The main call to action is to prioritize having an inclusive design mindset, aiming not only for safety and operational efficiency but for accessibility and usability for as many people as possible. This involves applying principles like Universal or Inclusive Design, depending on the context. It also emphasizes deeper engagement with participatory ergonomics, particularly during the design stages. By incorporating diverse perspectives, we ensure our designs remain user-centered and proactive in addressing potential challenges. This approach applies not only to physical design but also to processes, methods, and communications throughout the worker experience.

Know Your Contacts

You need to be "in the know" about broader organizational (or site-level) initiatives that could have an ergonomics impact. As such, build connections with key individuals in your organization who oversee DEI efforts (or employee relations), as well as get to know the various employee resource groups (ERGs). Take the time to meet with these contacts and discuss your goals for aligning ergonomics with their current and future diversity, inclusion, and equity initiatives. Together, you can identify actionable steps for your involvement in ongoing or upcoming initiatives.

Another critical contact is Innovation. This can look different in every organization, but from an ergonomics process perspective, you want to be in contact with the person or group (or groups) that are spearheading strategic and transformative projects and initiatives, including continuous process improvements and new business ventures. Ensure they are aware of your processes' objectives, goals, and capabilities—including inroads you have established within the organization through the participatory ergonomics program, which will be an important advantage if the organization is pursuing "intrapreneurship" programs. As noted in Chapter 4, this could be a mutually advantageous collaboration for ergonomics, DEI, and innovation efforts.

Now that you are on the contact list for organizational initiatives (such as pre-hire functional testing, innovation campaigns, automation projects, or life-work policies addressing issues like shift work, overtime, flexible work arrangements, etc.), be prepared, especially in terms of ensuring resources/time are available and incorporated into the structure of your process, to provide technical expertise. Offering ergonomics guidance, facilitating participatory feedback, or positing critical questions for consideration (e.g., the impact of fatigue management initiatives for various cultures) in these areas will help to ensure that the organization adopts efficient and effective policies that will meet the diverse physical and psychosocial needs of the workforce.

Practical Ways to Up Your Ergonomics Process Game

The following are a few quick considerations to undertake now to enhance your organization's ergonomics process:

- If holding targeted solution development events (or kaizen events) isn't part of your process, consider integrating them. I find these events to be an effective way to engage new "ergo converts," including skeptics ("Devil's Advocates") who may turn their doubts into constructive input. These events also build operational credibility by focusing on real, pressing issues. And the management of these events can serve as valuable leading indicators for ergonomics and DEI metrics (e.g., count of events, employee participation rates, number of solutions identified/implemented, return on investments of those solutions, etc.).
- And speaking of metrics, review your current ergonomics process metrics and identify at least one DEI-related metric to incorporate into your dashboard. Another action to consider is conducting benchmarking analytics, such as analyzing your organization's past data to correlate high-demand or high-risk positions with employee surveys, turnover data, and other relevant factors to identify any disparity gaps, then track progress from there.
- With regard to the scope of your risk screening, assessment, and management process, streamlining the incorporation of psychosocial risks into your process (in accordance with your organization's safety management system (e.g., ISO 45001/ISO 45003)), if not included already, will be a critical step for a more holistic and inclusive ergonomics approach. This could involve expanding screening assessments and process/equipment ergonomics design criteria to better capture psychosocial risks through the utilization of tools like the Copenhagen Psychosocial Questionnaire.
- Additionally, leaning into the participatory elements of your process to engage employees, ensuring that diverse needs are considered in risk assessments and mitigations, as well as part of any Prevention Through Design efforts, will help to prevent or minimize psychosocial risks and further promote a safety culture that prioritizes psychological safety. In addition to adding this to your risk assessment process, understanding and advocating for existing psychological safety initiatives, such as Employee Assistance Programs and life-work balance policies, is important. Wherever possible, integrate these programs throughout the ergonomics process, such as in ergonomics communications and training efforts. This will help to ensure that cognitive health and safety is prioritized alongside physical health and safety measures.
- And speaking of risk assessments, consider incorporating the Physical Demands Analysis (PDA) process into your risk screening process. As discussed in Chapter 7, the PDA process, which comprehensively evaluates the physical demands in every step of a process, provides a framework that is valuable for risk screening and pinpointing specific tasks for prioritization.

Look for Training Opportunities

Look for training opportunities to enhance both your personal skill set and your organization's ability to implement diverse and innovative ergonomics solutions. For example, to enhance your approach to DEI, consider exploring training opportunities that focus on DEI stewardship, such as enrolling in local DEI leadership programs (like the Riley Diversity Leadership Institute at Furman University mentioned in the Opening Introduction). Additionally, connect with your organization's training and development teams to explore diversity-related courses, including those on unconscious bias and empathetic communication, to increase awareness and improve interpersonal interactions. Beyond DEI, it is also valuable to pursue innovation-focused training, such as Design Thinking and courses on Universal Design and Inclusive Design, which will encourage creative, inclusive ergonomics solutions.

Ensure a Stronger Presence in Accommodations and Return-to-Work Processes

The next call to action for ergonomics practitioners is to ensure your/the ergonomics process's engagement in your organization's Accommodations and Return-to-Work programs. As with the other organization-wide initiatives noted earlier, these programs may be managed through various departments within the organization. And while it could pose a resource challenge for your ergonomics process, it is crucial for ergonomics to be integrated into these programs. At the very least, ergonomics expertise should be consulted to provide guidance on an as-needed basis to ensure the safe and effective reintegration of employees, addressing both physical and psychological needs.

Prepare an "Ergo Elevator Pitch"

The last call to action for ergonomics practitioners is to prepare a short (30 seconds–2 minutes; preferably the shorter the better) "elevator pitch" about the ergonomics process. In fact, it is always good to arm yourself with a quick rundown of what you do for the organization.

It would be good to tailor this pitch to your audience—e.g., when meeting with the DEI Officer/ERGs, include the specific "what's in it for them" in the presentation. And by presentation, keep in mind the pitch is not meant to be a slide deck, but a quick conversation detailing the key points: who you are (or what the ergonomics process is), what you do (or what the ergonomics process does), and how they (the audience) would benefit from you/the ergonomics process. It is also recommended that there is a good hook… I'm thinking:

> *Did you know that without ergonomics we can't meet our business goals? Here's how you can fix that!*

There are several guides available on preparing and delivering a good pitch (check with your local training and development department or the group managing intrapreneurship activities for any available training opportunities).

Calls to Action: DEI Practitioners/ERGs (or Advocates)

One immediate call to action is to connect with the ergonomics practitioner(s) within your organization. Take the initiative to engage with them (maybe share this book as a conversation starter) on how ergonomics can play a vital role in supporting DEI efforts and vice versa. Look for opportunities to identify and collaborate on key opportunities where ergonomics activities and objectives align with DEI goals.

Lastly, consider attending major professional ergonomics conferences like the Applied Ergonomics Conference, HFES, International Ergonomics Association meetings, or the National Ergo Expo. Connecting with ergonomics professionals in various domains and industries can provide invaluable insights and networking opportunities that could lead to ideas that further enhance your DEI efforts. Plus, these conferences often feature sessions on topics such as neurodiversity research, psychological safety, and practical approaches to making workplaces more inclusive. Engaging with these events will allow you to stay informed about the latest trends, research, and practical solutions that can directly benefit your DEI initiatives.

REFERENCES

Dul, J., Bruder, R., Buckle, P., Carayon, P., Falzon, P., Marras, W. S., & van der Doelen, B. (2012). A strategy for human factors/ergonomics: Developing the discipline and profession. *Ergonomics, 55*(4), 377–395. https://doi.org/10.1080/00140139.2012.661087

Hoffmeister, K., Gibbons, A., Schwatka, N., & Rosecrance, J. (2015). Ergonomics climate assessment: A measure of operational performance and employee well-being. *Applied Ergonomics, 50*, 160–169. https://doi.org/10.1016/j.apergo.2015.03.011

Lööw, J. (2019). Attractive work and ergonomics: designing attractive work systems. *Theoretical Issues in Ergonomics Science, 21*(4), 442–462. https://doi.org/10.1080/1463922X.2019.1694728

Mulaomerovic, E., & Wang, E. (2023). Challenges and opportunities for human factors/ergonomics as a strategic partner for business managers: In-depth research of experts' visions. *Sustainability, 15*(4), 3078. https://doi.org/10.3390/su15043078

Wooldridge, A. R., Grant, C., Widdowson, A., Rogers, C. C., & Figueroa Jacinto, R. (2022). Developing the human factors/ergonomics discipline to support diversity, equity and inclusion. *Proceedings of the Human Factors and Ergonomics Society Annual Meeting, 66*(1), 139–141. https://doi.org/10.1177/1071181322661020

Acknowledgments

Many thanks to my sisters and brother for their patience and love.

Index

Note: page numbers in **bold** refer to tables and those in *italics* refer to figures.

Kolibri 360 GmbH
Ferdinand-Jühlke-Straße 7
99095 Erfurt, Deutschland
produktsicherheit@kolibri360.de